ELECTRICAL ENGINEERING DEVELOPMENTS

PROVIDING RELIABLE OPERATION OF ELECTRIC NETWORKS

ELECTRICAL ENGINEERING DEVELOPMENTS

Additional books and e-books in this series can be found on Nova's website under the Series tab.

ELECTRICAL ENGINEERING DEVELOPMENTS

PROVIDING RELIABLE OPERATION OF ELECTRIC NETWORKS

ALEXANDER YU. KHRENNIKOV
ALEXEY A. KUVSHINOV
AND
IGOR A. SHKUROPAT

nova
science publishers
New York

NOTICE TO THE READER

The Publisher has taken reasonable care in the preparation of this book, but makes no expressed or implied warranty of any kind and assumes no responsibility for any errors or omissions. No liability is assumed for incidental or consequential damages in connection with or arising out of information contained in this book. The Publisher shall not be liable for any special, consequential, or exemplary damages resulting, in whole or in part, from the readers' use of, or reliance upon, this material. Any parts of this book based on government reports are so indicated and copyright is claimed for those parts to the extent applicable to compilations of such works.

Independent verification should be sought for any data, advice or recommendations contained in this book. In addition, no responsibility is assumed by the Publisher for any injury and/or damage to persons or property arising from any methods, products, instructions, ideas or otherwise contained in this publication.

This publication is designed to provide accurate and authoritative information with regard to the subject matter covered herein. It is sold with the clear understanding that the Publisher is not engaged in rendering legal or any other professional services. If legal or any other expert assistance is required, the services of a competent person should be sought. FROM A DECLARATION OF PARTICIPANTS JOINTLY ADOPTED BY A COMMITTEE OF THE AMERICAN BAR ASSOCIATION AND A COMMITTEE OF PUBLISHERS.

Additional color graphics may be available in the e-book version of this book.

Library of Congress Cataloging-in-Publication Data

ISBN: 978-1-53615-422-1

Published by Nova Science Publishers, Inc. † New York

CONTENTS

PREFACE

This book presents theoretical and practical aspects of providing reliable operation of electric networks by 6 – 500 кV voltage.

Chapter 1 shows that a serious problem of the glacial deposits of overhead transmission line's wires during the autumn-winter period. One of the traditional active methods is the melting of glacial deposits on alternating current lines by creating short circuits or direct current using uncontrolled or controlled rectifier blocks.

Chapter 2 deals with the expansion of the Automatic Dispatch Control System for substations and electrical networks using intelligent subsystems based on the technology of expert systems.

Chapter 3 presents an application's experience of monitoring of the electric activity of the partial discharges into the insulation of transformers, data of electrical tests in autotransformer windings and testing XLPE cables.

Chapter 4 presents an experience of thermal and electrodynamic tests of reactors on the reliability and durability of windings during the passage of short-circuit currents serve as a tool for improving the reliability of their design.

Investigation of Technological breakdowns with the damage of power transformers is an important difficult task and is performed to improve the reliability of power supply in chapter 5.

Chapter 6 shows diagnostic and mathematical models of the windings of the high-voltage transformers and reactors. Electromagnetic forces acting on the windings of a two-winding transformer are also considered. The case of finding the current-flowing conductors of the transformer in a magnetic leakage field the electrodynamic forces act on them.Mathematical models of winding radial and axial deformations are received.

A comparative characteristic of the switching capabilities of photothyristors, lockable thyristors IGCT and high-power transistors IGBT is given, taking into account the specific operating conditions in the composition of the high-voltage high-current semiconductor key (HHCSK) during electrodynamic tests in chapter 7. A safety criterion for electrodynamic tests is proposed, which consists in the forced cessation of the short circuit experience immediately after the formation of an emergency state signal for the windings of the tested power transformer.

The authors consider the basic operations of implicative algebra for the choice of the logical-algebraic modeling of static converters, provided that the objective and predicate variables are identified with the parameters of the energy and information processes in chapter 8.

Chapter 9 presents controlled reactor-transformers with reduced high current harmonics and calculation of parameters of active parts of transformers.

The results of designing device for calculating current on digital current transformers, based on the Faraday Effect, are in the chapter 10. The device calculates actual value of current, writes data to the non-volatile memory and sends status messages to the automated workstation. During design the component base of developed device was selected, metrological calculation was made, circuit diagram and printed circuit board of the devise was designed, and algorithms of the device was developed.

The book is based on the previous author's researches. The author has more than 240 scientific and technical publications. Book is intended for the leaders and the specialists of the technical service enterprises of electrical and distribution networks and electrical stations, students of higher educational universities, students of the courses of an increase in the qualification of power engineers, researchers & scientists in the field of electrical engineering.

A. Yu. Khrennikov, PhD
A. A. Kuvshinov, PhD
and
I. A. Shkuropat

Chapter 1

A Review of Modern Semiconductor Systems for De-Icing Overhead Transmission Line Wires

A. A. Kuvshinov, PhD and A. Yu. Khrennikov*†*, PhD*

Abstract

The chapter shows that when operating overhead transmission lines in a number of regions, there is a serious problem of the glacial deposits of wires during the autumn-winter period. As a passive measure against the glacial deposits, various wires of increased strength can be used. One of the traditional active methods is the melting of glacial deposits on alternating current lines by creating short circuits or direct current using uncontrolled or controlled rectifier blocks.

The development of new means to prevent glacial deposits on the overhead transmission lines consists of the use of combined conversion units capable of performing melting of glacial deposits, if necessary, and the rest of the time compensating for reactive power.

The most promising one should recognize the melting of glacial deposits with an ultra-low frequency current that combines the advantages of melting with an alternating current of the industrial frequency (on three

* Corresponding Author's E-mail: alekseikuvshinov@yandex.ru.

† Author's E-mail: ak2390@inbox.ru.

wires at the same time) and a DC current (limited only by the active resistance, smooth regulation of the melting current).

Keywords: overhead transmission lines, melting of glacial deposits, direct current, ultra low frequency current

1.1. INTRODUCTION

In a number of regions, operation of overhead transmission lines is severely complicated by ice formation of the wires in autumn and winter, due to the fact that average time of eliminating icing-related emergencies exceeds average time of eliminating emergencies caused by other reasons by 10 and more times. Research has shown that ice coating on overhead line wires is formed at the air temperature of approx. -5 C° and wind speed of 5...10 m/s. Permissible thickness of ice coating is from 5 mm to 20 mm for overhead lines with nominal voltage of (3 – 330) kV, located in climatic regions of I to IV ice formation severity [1].

Various high-strength cables/wires can be used as a passive de-icing method. For examples, ACCC wire (Aluminum Conductor Composite Core wire), which is a number of aluminum wires placed around carbon-fiber and fiberglass epoxy core [2]. ACCC wire core keeps stable dimensions because its thermal expansion coefficient ($1.6 \cdot 10^{-6o}C$) is almost ten times less than steel has ($11.5 \cdot 10^{-6o}C^{-1}$). Due to this, ACCC wires can withstand high temperature for a long time, thus preventing ice coating formation.

It is also worth noting Aero-Z® wire, consisting of one or several concentric wires (internal layers) and Z-shaped wires (external layers). Each layer of the wire has lengthwise twist of a definite pitch. Smooth surface reduces wind load by 30-35% and prevents snow and ice buildup. However, Aero-Z® wire has limitation for ice melting time because it cannot withstand temperature increase over 80°C for a long time.

Overall, practical implementation of passive de-icing methods is possible only in cases of designing and commissioning new transmission lines. Modernization of ‘old’ overhead lines would incur significant costs.

Due to these reasons, research for methods of active de-icing of overhead lines is urgent. Among traditional methods, one can note melting ice on overhead line wires with alternative current by artificially inducing short circuits or with direct current by using uncontrollable or controllable rectifiers [3, 4]. However, in the first case, it is possible to damage overhead line wires, and in the second case, the expensive rectifiers are not used for most part of the year. At the same time, modern hardware of power electronics provides additional opportunities and incentives to develop new de-icing methods devoid of the indicated drawbacks. Issues of ice coating formation and de-icing are topics of a large number of scientific papers. This work sets the task to systemize and perform comparative analysis of existing methods of de-icing, and finding solution to ice coating problem will help choose the most suitable one for particular local conditions from the many technical solutions at hand.

1.2. Classification of De-Icing Methods

The known de-icing devices and methods use the following types of physical impact to remove ice and rime deposits from transmission line wires (Figure 1.1):

- thermal impact by heating the cable to the temperature of $(120 - 130)^0$C, which melts the ice coating, or by preventive wire heating to $(10 - 20)^0$C to prevent ice coat formation;
- thermodynamic by preliminary heating until a melt layer is formed between the wire and the ice coating and subsequent “shaking up” of the wires with by Ampere force, arising when a powerful current pulse passes;
- electromechanical impact by creating cyclical current pulses that induce mechanic vibration of the wires cause ice coating destruction. Efficiency of electromechanical impact is enhanced when such current pulses parameters are used that induce mechanical resonance;

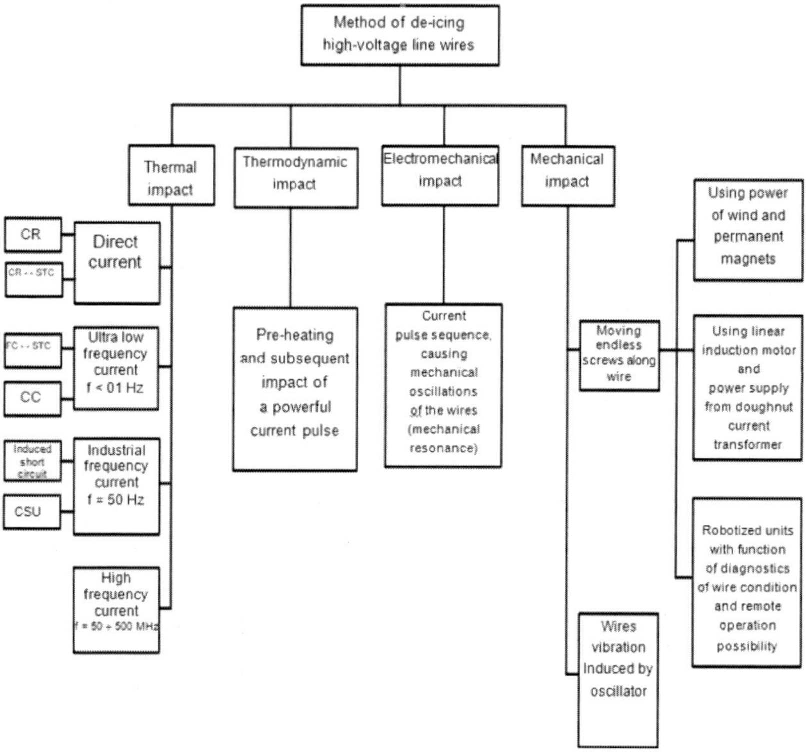

Figure 1.1. Classification of known methods of removing ice coating from overhead line wires (CR – controllable rectifier; STC – static thyristor compensator; FC – frequency converter; CC – cycloconverter; SCU – series compensation unit).

- mechanical impact by moving endless screws along the wire, using wind energy, energy of high-voltage line phase current electromagnetic field, permanent magnets, linear induction motor, by inducing wire vibrations with the help of an oscillator (these shall not be further considered because they do not use converter equipment). We shall consider a common drawback of mechanical systems, which is the need for manual mounting on the wire, dismounting from the wire, as well as moving from one wire to another. This requires special technical means (vehicle-mounted aerial work platform) and maintenance personnel, which increases operating expenses and complicates operation in hard-to-reach areas.

1.3. Thermal Impact with Alternate Current of Industrial Frequency

Melting ice with alternate current is used only on the lines of voltage lower than 220 kV, with wires cross-section less than 240 mm^2 [3]. Power is supplied, as a rule, by substation buses of voltage of 6 - 10 kV, or by a separate transformer unit. Ice melting circuit should be selected in such a way as to ensure that the current passing through high-voltage lines exceeds permissible continuous current by 1.5 – 2 times. Such excess is justified by the short duration of melting process (~1 h), and by more effective cooling of the wire in wintertime. For standard steel reinforced aluminum Type AC wires with cross-section of (50 ÷ 185) mm, the estimate value of one-hour ice melting current intensity lies in the range of (270 ÷ 600) A, while the intensity of the current preventing wire icing lies in the range of (160 ÷ 375) A.

However, it is impossible to select the necessary value of short circuit current only by selecting ice melting circuit. Exceeding the above-mentioned values of melting current can cause wires burnback and subsequent irreversible loss of strength. At lower values, a single passing of short circuit current may not be sufficient to fully remove the ice coating. Then, short circuits have to be repeatedly induces, which further worsens the consequences.

It is possible to avoid the mentioned negative consequences by using an alternate current thyristor compensator, the scheme of which is shown in Figure 1.2 [5]. In ice melting mode, switch 7 is OFF, and switch 8 is ON. Possible methods to regulate melting current are pulse-phase method by changing switching angles of power thyristors 1, 2, and 3; or pulse width method, by changing number of voltage supply periods.

In reactive power compensation mode, switch 7 is ON, and switch 8 is OFF. In this case, power thyristors 1, 2, and 3 and reactors 4, 5, and 6 form a thyristor-reactor group connected into a triangle, that is a part of the static thyristor compensator. The authors consider the possibility to use capacitors instead of reactors. In this case, reactive power compensation will be performed with the help of a regulated capacitor bank.

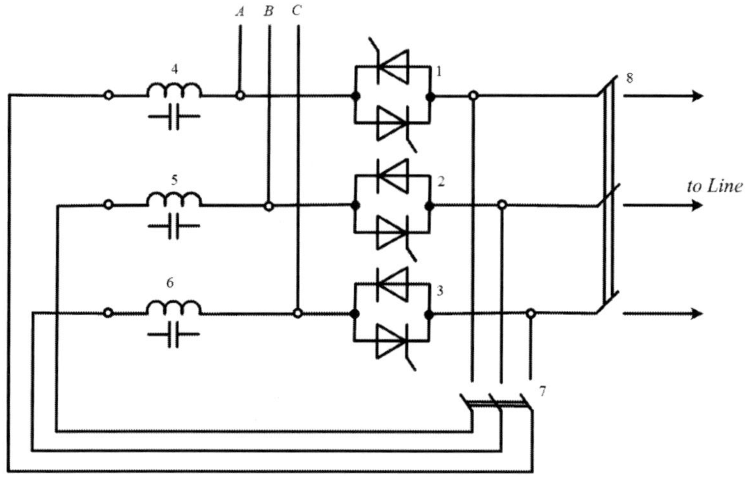

Figure 1.2. Unit for compensation of reactive power and ice melting.

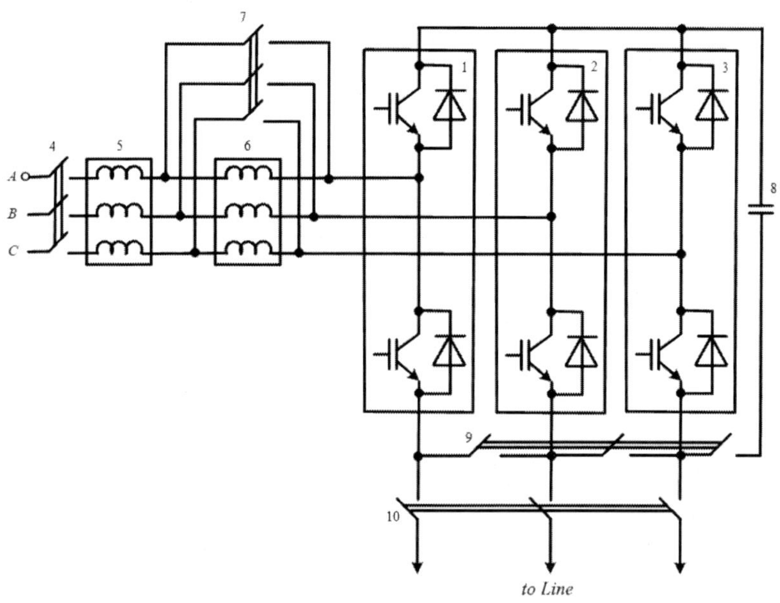

Figure 1.3. A combined unit for compensation of reactive power and ice melting.

However, whichever regulation method is used, ice melting is done by alternate current of industrial frequency and requires significant power supply (dozens of MVA), because real resistance of overhead line wires is much less than inductive resistance. Total supply power is increased due to large reactive load which is useless for ice melting. It is possible to increase melting efficiency by applying series capacitive compensation of inductive resistance, if capacitors are used as part of the proposed unit. However, the authors did not consider such alternative.

It is also worth considering a combined unit for compensation of reactive power and ice melting, the scheme of which is shown in Figure 1.3 [6]. In ice melting mode, switch 7 is ON and shunts reactor 6, switch 9 shuts off capacitor bank 8, and switch 10 is ON. This enables simultaneous melting on all wires of overhead line.

In of reactive power compensation mode, switches 7 and 10 are OFF, and switch 9 is ON. The result is a typical static compensator circuit consisting of transistor modules 1, 2 and 3, reactors 5 and 6 on alternate current side, and capacitor bank 8 on direct current side. This structure can work in power generation mode as well as in reactive power consumption mode.

The unit shown in Figure 1.3 has a significant drawback, in particular, incomplete use of gating section in melting mode. This can be explained by the fact that the melting current passes only through the 'lower' phase switches 1, 2 and 3 of the converter bridge. Converting the bridge circuit into three alternate current switches will require additional switching gear and a substantially more complicated power circuit.

1.4. Thermal Impact of Direct Current

For the first time, ice melting with direct current as a prospective method of de-icing high-voltage phase wires was mentioned in [7]. Among the first mass produced units for direct current ice melting were silicon cell ice melting rectifiers VUKN-16800 - 14000, created using Larionov circuit, based on silicone uncontrollable rectifiers VK-200 with rectified voltage of

14 kV, rectified current of 1200 A, and output power of 16800 kW [8]. Schemes of ice melting with rectified current are considered in detail in [4].

Main drawbacks are that high-voltage lines must be shut off, while rectifier is not operated for most part of the year because there is a necessity for ice melting only in winter. It is worth noting the proposal to melt ice by pulsing current without shutting off high-voltage lines [9]. Rectifier is connected to the cut of the heated wire in such a way that direct current does not pass through power transformer and current transformer. Wires are heated with alternate current with alternating-current component that is defined by high-voltage line load, and direct-current component defined by rectified voltage and real resistance of melting circuit. However, this proposal does not increase rate of using rectifiers and will require additional switching gear for practical implementation.

This justifies attempts to extend functional capabilities by combining in one unit an ice melting rectifier and a reactive power compensation unit. This enables all-year operation, which significantly increases economic efficiency.

JSC NIIPT, Scientific & Research Institute of Direct Current (Saint-Petersburg) has developed a container-type converting unit for the combined unit for compensation of reactive power and ice melting (see Figure 1.4) [10].

The converting unit (Figure 1.4) includes transporting container 1, thyristor units 2, and control modules 3, forced air cooling system 4, disconnector 5 with electric power drive 6, anode lead 7, cathode lead 8, phase lead 9 of converter bridge, Control, Regulation, Protection and Automation System (SRPAS) 10, disconnectors 11, and 12, and capacitor tanks 13.1, 13.2, and 13.3.

Power equipment is designed for operation in moderate to cold climate zones (design for UHL1 climate zone, acc. to Russian GOST) and placed into a closed steel container set upon a basement in an open part of the substation. Power supplied is delivered from 10 kV coil from a dedicated transformer. The converting units shown in Figure 1.4a are used to make a combined unit, shown in Figure 1.4b.

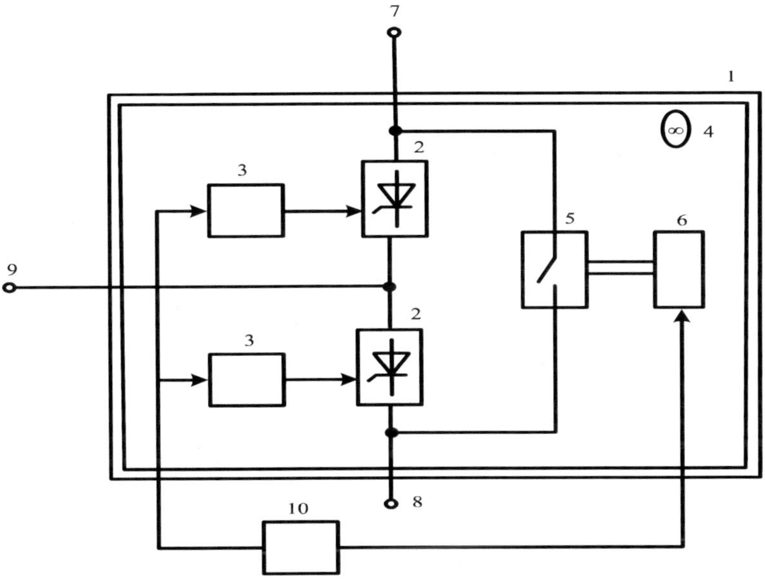

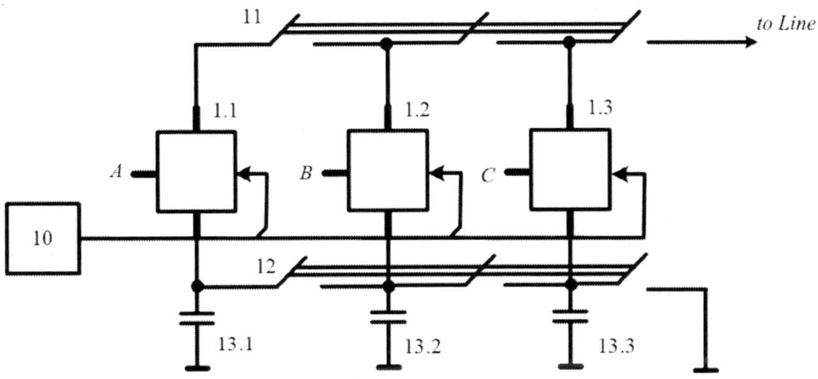

Figure 1.4. Scheme of container-type converting unit (a) and combined unit (b) for compensation of reactive power and ice melting, disconnectors 11, and 12, and capacitor tanks 13.1, 13.2, and 13.3.

In ice melting mode, disconnectors 11 and 12 are closed (see Figure 1.4b), disconnectors 5 (see Figure 1.4a) are open. A circuit of a three-phase bridge rectifier is created that provides nominal rectified voltage of 14 kV, nominal melting current of 1400 A, and capability to regulate melting current in the range of (200...1400) A.

In reactive power compensation mode, disconnectors 11 are 12 open, and disconnectors 5 are closed. This creates a circuit of capacitor tanks13.1, 13.2, and 13.3, which is controlled by thyristor modules aligned in parallel-opposite connection 2. However, compensation mode allows only step control of reactive power.

It is possible to avoid the latter drawback by using a combined unit for ice melting and reactive power compensation, shown in Figure 1.5 developed by JSC NIIPT, Scientific & Research Institute of Direct Current [11].

The combined unit for compensation of reactive power and ice melting consists of feeding transformer 1, three-phase disconnectors 2, and 16, three-phase reactors 3, and 15, high-voltage bridge converter 4, direct current capacitor tank 5, one-phase disconnectors 6, and 7, control system 8, assemblies 9 ÷ 14 of fully controllable devices with reverse diodes, and resonance transformer 17.

In ice melting mode, disconnectors 6, 7 and 16 are ON. Melting is done with direct current. Melting current is regulated by high-voltage pulse width modulation method (PWM). For example, when load current passes through diodes of assemblies 13 and 10, the fully controllable device from assembly 9 or 14 connects in PWM mode. This causes a short-time loop of double-phase short circuit 9 – 10 or 13 – 14. The load is shunted, and melting current is regulated. Speed of short-circuit current buildup is limited by reactor 3. Due to selecting PWM frequency and rate, the thyristor is blocked before short-circuit current increases to a dangerous value. At the same time, conducting interval of thyristor is less than in reactive power compensation mode. In reactive power compensation mode, disconnectors 6, 7 and 16 are OFF. High-voltage bridge converter 4 operates in STATCOM (static compensator) mode.

A number of authors who judge by their own work experience are of opinion that only 7% to 30% of the heated wire length is actually coated in ice during the melting. This is explained by the fact that different areas of high-voltage line happen to be in different climatic conditions due to rotation angle and impossibility to predict wind direction at the moment of ice coating formation. Consequently, significant amount of electric power is wasted. Therefore, a mobile unit is proposed that will enable to go to those areas of high-voltage lines where ice coating on the wires is identified.

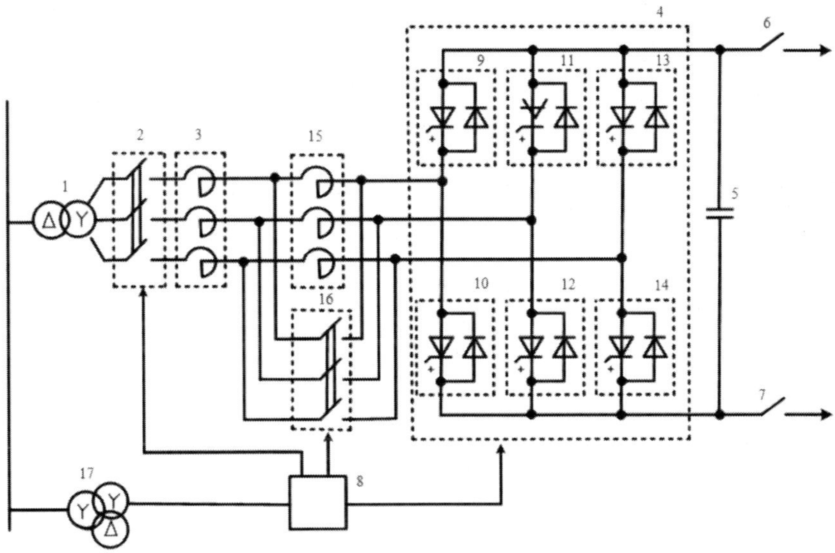

Figure 1.5. Combined unit for compensation of reactive power and ice melting.

Mobile generator for melting ice on overhead transmission lines [12] is installed on vehicle-mounted platform, the three-phase bridge rectifier is powered (0.4 kV) by two diesel ADV320 generators, each generating 320 kW. It is equipped with conductors and terminals to connect to overhead line wires, and with electric buses to connect wires in the span between the towers into ice melting circuit. The proposed technical solution enables ice melting on the length of two spans of overhead line on phase conductions and ground wire.

Common drawback of all devices that use direct current to create thermal impact is the necessity to use ice melting scheme "wire - wire" or "wire - two wires." Both schemes lead to increase of melting time and, consequently, power costs. To decrease melting time, melting scheme "three wires - earth" should be preferred, but grounding devices of substations, as a rule, are not designed to hold relatively long passing of direct current of strength up to 2000 A.

1.5. Thermal Impact by Ultra-Low Frequency Current

Technical meaning of this impact is that melting is done by low-frequency current formed by an autonomous three-phase voltage inverter, while effective value of melting current is created and maintained at the required level by changing supply voltage [13-18].

At output voltage frequency of autonomous inverter of tenths of Hz or less, current strength in line wires is limited basically only by real resistance. This lets increase permissible length of overhead line as compared to the process of melting by industrial-frequency alternate current, simplifies melting organization, reduces ice melting time, and decreases the number of additional switching gear.

Scheme of a combined unit for compensation of reactive power and ice melting that implements the proposed solution is shown in Figure 1.6.

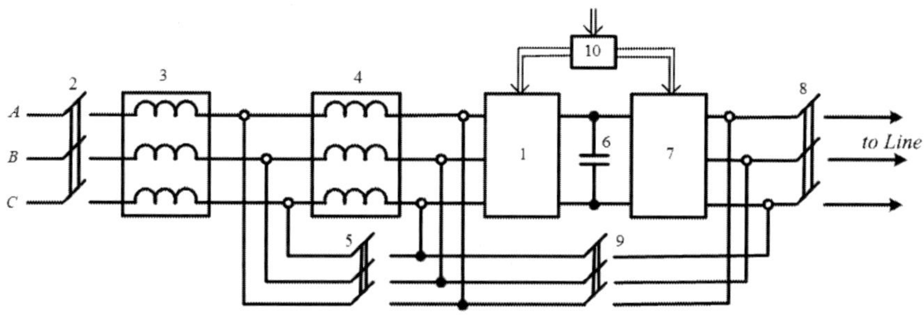

Figure 1.6. A combined unit for compensation of reactive power and ice melting.

The combined unit is made up of three-phase bridge inverters on fully controllable semiconductor switches 1 and 7, triple-pole switches 2, 5, 8, and 9, three-phase choke coils 3, and 4, capacitor tank 6, and control system 10.

In ice melting mode, switches 5 are 8 ON, and switch 9 is OFF. Bridge converter 1 works as a controllable rectifier, and bridge converter 7 works as an autonomous three-phase voltage inverter. Melting is performed on three wires of the overhead line simultaneously.

In reactive power compensation mode, switches 5 and 8 are OFF, and switch 9 is ON. Bridge converters 1 and 7 work in parallel.

Switching angle α is selected as a little less than 180 electric degrees. Low real power is consumed from the network, necessary to maintain voltage in capacitor tank 6. Alternate voltage is formed on the side of alternate current of bridge inverters 1 and 7, and its first harmonics is shifted in relation to power source phase voltage by angle $\beta = (180 - \alpha)$. If amplitude of first harmonics of the produced voltage exceeds that of power source voltage, then bridge converters 1 and 7 generate reactive power, otherwise they consume reactive power. Changing high-frequency PWM rate regulates first harmonics amplitude of the formed voltage, and, consequently, value and direction of reactive power.

1.6. Thermal Impact of High-Frequency Current

Thermal impact of high-frequency current is that, without shutting off overhead lines from consumers, generator supplies high-frequency current (50 to 500) MHz via interface unit and high-voltage coupling capacitors to phase conductors [14].

In homogenous conductor, alternate current is concentrated in surface layer, and thinning of this layer leads to increase of current-conducting part of the conductor. It means that with the same strength of the current passing through the wire, the higher the signal, the larger the thermal power dissipated through the conductor. Foe examples, at $f \geq 100$ MHz, resistance of aluminum wires increases by 600 times and more.

It has been shown that high-frequency generator power of several dozens of kW is enough to heat the wire to to $(10 \div 20)°C$, which should prevent formation of ice coating. To remove the already formed ice coating, it will be necessary to heat the wire not to $(10 \div 20)°C$, but to $(100 \div 180)°C$, in order to melt the ice. This will consequently require much more electric power and time.

Therefore, this method is the most efficient for preventing ice coating formation because it can be used without shutting off power from the consumers. However, use of generators working in frequency range of 87.5…108 MHz can cause high-intensity radio interference in FM band.

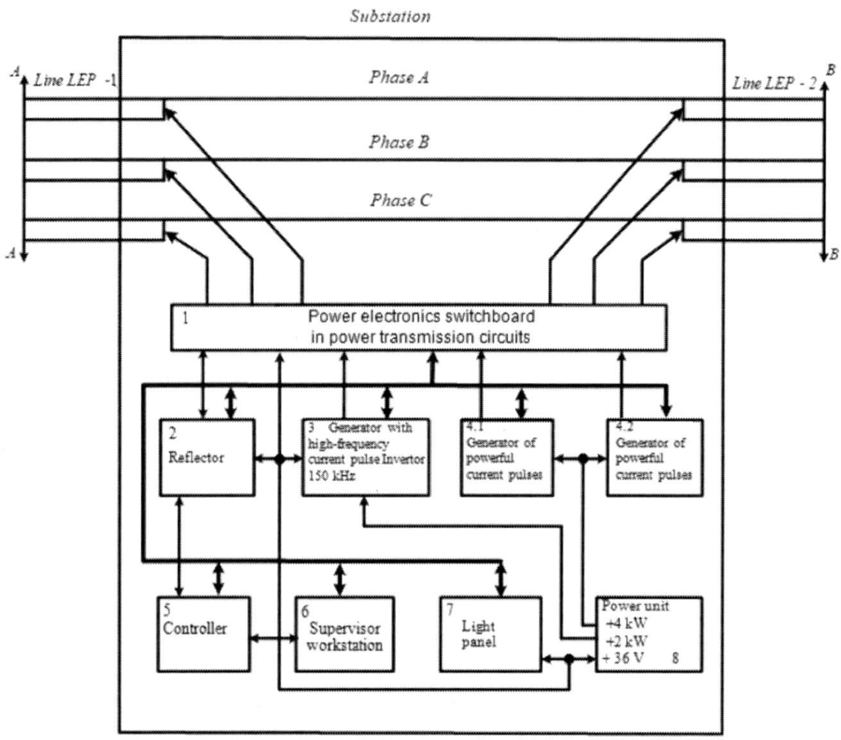

Figure 1.7. Device for removal of snow and ice coating from power transmission line wires.

1.7. Thermodynamic Impact

Heating the wires with high-frequency current not only prevents ice coating formation, but can also be used to facilitate removal of already formed ice coating. This method is used in the device whose scheme is shown in Figure 1.7 [15].

Pulse reflection meter 2 continuously senses power transmission lines LEP-1 and LEP-2 as regards total time of pulse signal direct and reverse propagation after the signal is reflected from end of the line. Propagation time depends on the ice coating formation and its properties.

High-frequency current pulse generator 3 pre-heats the wire and creates a melt layer between the wire and ice coating. The subsequent impact of generator current power pulse 4 creates a thermodynamic shock, which together with the pre-heated melt layer de-ices power transmission line wires.

Computer workstation (WKS) of supervisor 6 and controller 5 ensure uninterrupted operation of and display operational information on light panel 7.

1.8. Electromechanical Impact

It is known that parallel wires, when the current passes through them, attract or repel each other. At regular passing of current pulses, high-voltage line wires will oscillate mechanically, thus removing ice and rime deposits. Current pulses frequency should be close to mechanical resonance and of such amplitude as to overcome internal and external friction. Change of the passing current can be periodic, of swept frequency, harmonic, or shaped as a pulse packet with predetermined laws of change of frequency, amplitude and relative duration [16]. Figure 1.8 shows one of possible variants of automated system for de-icing implementing the proposed method.

Power transformer 1 transforms supply voltage to the required value. Power electronics unit rectifies voltage received from power transformer 1 and creates current pulses of required value, shape and frequency that pass through overhead line wires 2. The control system, which is a programmable logic controller, processes the data received from external sensors of ice and wind load 3, humidity 4, and temperature 5, defines required current pulse shape and frequency for power electronics unit, and controls the system in whole [20-28].

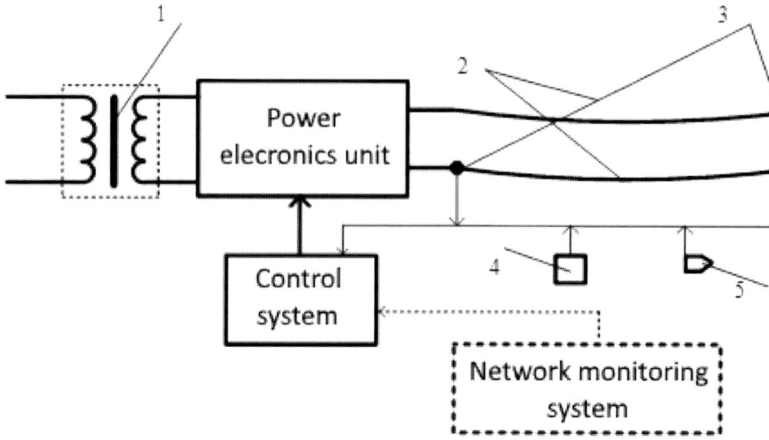

Figure 1.8. Scheme of electromechanical impact to overhead line wires for de-icing.

Practical implementation of this method requires a thorough calculation of current pulse value and frequency to avoid possible negative consequences of resonance. To increase efficiency of de-icing, current pulses should pass through wires lying on different levels. It would enable use of ice inertia and gravity as an additional destructive factor [29-37].

This method requires shutting off the overhead lines, the same as melting method. However, time required for mechanical de-icing is significantly less than time required for melting. Therefore, power cost for this de-icing method will be less than for ice coating melting method.

Conclusion

The dominant trend in development of new methods for de-icing high-voltage line wires is the use of combined converter units that can melt the ice when necessary and compensate reactive power during the rest of time. The most advantageous method is ice melting by ultra-low frequency current that combines advantages of melting by industrial frequency alternate current (on three wires simultaneously) and melting by direct current (limited only by real resistance; melting current can be smoothly regulated).

Another advantage of this method is that the unit for ice melting by ultra-low frequency current is easily transformed into a static compensator of reactive power. This lets use the costly converting equipment during the whole year. However, this method has the disadvantage of the necessity to shut off overhead lines for de-icing.

The latter disadvantage can be completely avoided by using technology of flexible alternate power transmission [18] that involves converting equipment theoretically capable, if necessary, to provide regular wire heating that prevents formation of ice coating.

Acknowledgments

The authors are grateful to the staff of the research and production company "Energy-T" for useful remarks and practical implementation of the equipment for melting ice with direct current on the wires of overhead lines of electric networks "Ulyanovskenergo" and "Bashkirenergo."

References

[1] Electrotechnical manual, in 3 volumes. Volume 3. In *2 folios. Folio 1. Generation and distribution of electric power* (Edited by MPEI Professors: Orlova, I. N. (Senior Editor) et al.). Version 7, amended and complemented. – Moscow: Energoatomizdat. – 1988. – 880p. (in Russian).

[2] Alekseev, B. A.. Improvement of overhead transmission lines capacity and use of wires of new grades. *ELEKTRO Journal*. – 2009. – Issue 3. – pp. 45-50. (in Russian).

[3] National Standard Guideline Document RD 34.20.511 (34-70-028-82) *Guidelines for ice melting by alternate current*. Part 1. Moscow: Soyuztekhenergo, 1983. (in Russian).

[4] National Standard Guideline Document RD 34.20.511 (34-70-028-82) *Guidelines for ice melting by direct current*. Part 2. Moscow: Soyuztekhenergo, 1983. (in Russian).

[5] Patent of Russian Federation (RF) № 2505898 МКИ H02G7/16, H02J3,18. *Device for compensation of reactive power and ice melting*/ Stashinov, Yu. P., Konopelko, V. V. – Publ. 27.01.2014.

[6] Patent of RF № 2505903 МКИ H02J3/18, H02G7/16. *Combined unit for compensation of reactive power and ice melting*. Stashinov, Yu. P., Konopelko, V. V. – Publ. 27.01.2014.

[7] Burgsdorf, V. V. Ice melting by direct current without shutting off the line. *Electrical stations*. – 1945. – Issue 11. (in Russian).

[8] High-voltage rectifying unit of type VUKN-16800-14000 (silicon cell ice melting rectifier). Annotated list of major research and experimental works performed in Mordovia Research Electrotechnical Institute (1965-1968). – *Informelektro*, 1970. (in Russian).

[9] Genrikh, G. A., Denisenko, G. I., Mishin, V. V., Stryapan, V. N.. *Special modes of powerful static converters of ice melting units on power transmission lines*. Kiev – Publishing union "Vishha shkola." – 1975. – 242p. (in Russian).

[10] Patent of RF № 2390895 МКИ H02G7/16, H02J3/18. *Converting container-type device for combined ice melting and reactive power compensation unit*. Gurevich, M. K., Kozlova, M. A., Lobanov, A. V., Repin, A. V., Shershnev, Yu. A.,. – Publ. 27.05.2010.

[11] Patent of RF № 2376692 МКИ H02G7/16, H02J3/18. *Combined unit for ice melting and reactive power compensation*. Gurevich, M.K., Repin, A. V., Shershnev, Yu. A. – Publ. 20.12.2009.

[12] Patent of RF № 2522423 МКИ H02G7|16. *Mobile generator for ice melting on overhead transmission lines*. Kozlov, A. V., Chulkov, A.N., Shurupov, A. B., Vinogradov, A. A. – Publ. 10.07.2014.

[13] Patent of RF № 2505897 МКИ H02G7/16. *Method of controllable alternate current ice melting on overhead transmission lines*. Stashinov, Yu. P., Konopelko, V. V. – Publ. 31.05.2012.

[14] Patent of RF № 2356148 МКИ H02G7/16. *Method and device for de-icing of power transmission lines*. Kaganov, V. I. – Publ. 20.05.2009.

[15] Patent of RF № 2520581 МКИ H02G7/16. *Device for removing ice and rime coating from power transmission line wires.* Shelkovnikov, N. D., Shelkovnikov, D. N. – Publ. 27.06.2014.

[16] Patent of RF № 2166826 МКИ H02G7/16, B60M1/12. *Method of de-icing wires of overhead contact system and powertrain transmission lines*. Efimov, A. V., Galkin, A. G. – Publ. 10.05.2001.

[17] Patent of RF № 93184 МКИ H02G7/16 for a useful model. *Device for de-icing of power transmission lines.* Sattarov, R. R., Ismagilov, F. R., Almaev, M. A. – Publ. 20.04.2010.

[18] Kochkin, V. I. New technologies to improve power transmission lines capacity. Controllable power transmission. *Electrotechnics News.* – 2007. – Issue 4(46). (in Russian).

[19] Li, P., Zhao, N., Zhou, D., Cao, M., Li, J., Shi, X. Multivariable Time Series Prediction for the Icing Process on Overhead Power Transmission Line. *Scientific World Journal,* Volume 2014, Article ID 256815, 9 p.

[20] Hou, H., Yin, X., Chen, Q., You, D., Tong, G., Shao, D. Review on the wide area blackout of 500 kV main power grid in some areas of south china in 2008 snow disaster. *Automation of Electric Power Systems*, vol. 32, no. 11, pp. 12–15, 2008.

[21] Savadjiev, K., Farzaneh, M. Analysis and interpretation of icing rate meter and load cell measurements on the Mt. Bélair icing site. In *Proceedings of the 9th International Offshore Polar Engineering Conference*, vol. 2, pp. 607–611, Brest, France, 1999.

[22] Huang, X. B., Sun, Q. D. Mechanical analysis on transmission line conductor icing and application of online monitoring system. *Automation of Electric Power Systems*, vol. 13, no. 14, pp. 98–101, 2007.

[23] Yang, L., Hao, Y., Li, W. A mechanical calculation model for on-line icing-monitoring system of overhead transmission lines. *Proceedings of the Chinese Society of Electrical Engineering*, vol. 30, no. 19, pp. 100–105, 2010.

[24] Xu, Y., Bosisio, R. G. Goubau ice sensor transitions for electric power lines. *Sensing and Imaging*, vol. 10, no. 1-2, pp. 31–40, 2009.

[25] Gu, Y. H., Sistiaga, U., Sonja, M. "Automatic surveillance and analysis of snow and ice coverage on electrical insulators of power transmission lines." In: *Proceedings of the International Conference (ICCVG '08), vol. 5337 of Lecture Notes in Computer Science*, pp. 368–379, Springer, 2008.

[26] Sakamoto, Y. "Snow accretion on overhead wires," *Philosophical Transactions of the Royal Society A*, vol. 358, no. 1776, pp. 2941–2970, 2000.

[27] Farzaneh, M. and Savadjiev, K., "Statistical analysis of field data for precipitation icing accretion on overhead power lines," *IEEE Transactions on Power Delivery*, vol. 20, no. 2, pp. 1080–1087, 2005.

[28] Makkonen, L. "Modeling power line icing in freezing precipitation," *Atmospheric Research*, vol. 46, no. 1-2, pp. 131–142, 1998.

[29] Wu, X., Li, L., and Rui, X. M. "Icing load accretion prognosis for power transmission line with modified hidden semi-Markov model," *IET Generation, Transmission & Distribution*, vol. 8, no. 3, pp. 480–485, 2014.

[30] Rajakarunakaran, S., Devaraj, D., Venkatratnam, G. S. and Rao, K. S. P. "Articial neural network approach for fault detection in LPG transfer system," *Applied Soft Computing*, vol. 8, no. 1, pp. 740–748, 2008.

[31] McComber, P., De Lafontaine, J., and Laflamme, J., "A neural system to estimate transmission line icing," in *Proceedings of the 8th International Workshop on Atmospheric Icing on Structures (IWAIS '98)*, pp. 101–106, Reykjavik, Iceland, 1998.

[32] Larouche, E., Rouat, J. and Bouchard, G. "Exploration of static and time dependent neural network techniques for the prediction of ice accretion on overhead line conductors." In: *Proceedings of the 4th International Workshop on Atmospheric Icing of Structures (IWAIS '00)*, Chester, UK, 2000.

[33] Li, P., Li, N., Li, Q., Cao, M., and Chen, H. "Prediction model for power transmission line icing load based on data-driven," *Advanced Materials Research,* vol. 144, pp. 1295–1299, 2011.

[34] Li, P., Li, Q., Cao, M., Gao, S., and Huang, H. "Time series prediction for icing process of overhead power transmission line based on BP neural networks." In: *Proceedings of the 30th Chinese Control Conference (CCC '11)*, pp. 5315–5318, Yantai, China, July 2011.

[35] Li, Q., Li, P., Zhang, Q., Ren, W., Cao, M. and Gao, S. "Icing load prediction for overhead power lines based on SVM." In: *Proceedings of the International Conference on Modelling, Identification and Control (ICMIC '11)*, vol. 6, pp. 104–108, June 2011.

[36] Makkonen, L. "Models for the growth of rime, glaze, icicles and wet snow on structures," *Philosophical Transactions of the Royal Society A: Mathematical, Physical and Engineering Sciences*, vol. 358, no. 1776, pp. 2913–2939, 2000.

[37] Kollár, L. E., Farzaneh, M., and van Dyke, P. "Modeling ice shedding propagation on transmission lines with or without interphase spacers," *IEEE Transactions on Power Delivery*, vol. 28, no. 1, pp. 261–267, 2013.

Chapter 2

COMPUTER SUPPORT OF DISPATCHING SOLUTIONS IN POWER SYSTEMS BASED ON TECHNOLOGY OF EXPERT SYSTEMS

A. Yu. Khrennikov*, PhD

ABSTRACT

The chapter deals with the expansion of the functions of the Automatic Dispatch Control System for substations and electrical networks of industrial enterprises using intelligent subsystems based on the technology of expert systems. Intelligent subsystems contain semantic description of topology control objects and special programs-reasoning based on the technological instructions. An adviser for the operational and supervisory staff on analysis of contingencies, search planning system damage in distributive electric networks were reviewed. A new Simulator-Simulator function analysis of contingencies, providing automated checking skills dispatch personnel analyze situations related to technological injury in electrical networks of industrial enterprises.

* Corresponding Author's E-mail: ak2390@inbox.ru.

Keywords: dispatching management of electrical networks of industrial enterprises, expert systems, analysis of contingencies, disaster recovery, expert systems

2.1. INTRODUCTION

The article deals with the expansion of the functions of the Automatic Dispatch Control System for substations and electrical networks of industrial enterprises using intelligent subsystems based on the technology of expert systems. Intelligent subsystems contain semantic description of topology control objects and special programs-reasoning based on the technological instructions. An adviser for the operational and supervisory staff on analysis of contingencies, search planning system damage in distributive electric networks were reviewed. A new Simulator-Simulator function analysis of contingencies, providing automated checking skills dispatch personnel analyze situations related to technological injury in electrical networks of industrial enterprises.

In the last few years (2013-2017), the technical literature published mainly results on automation of operational dispatch analysis of emergency (including emergency) situations in the electrical networks of industrial enterprises [1-9].

It is necessary to systematize these works within the framework of a single concept. Analysis of abnormal situations can be conditionally divided into at least two main directions:

- a complete “relay” analysis of the situation, performed by specialists of relay protection in order to fix possible malfunctions of relay protection and automation devices,
- an operational dispatch analysis of the situation for duty personnel (conditionally this type of analysis will be called dispatching express analysis).

Instructions for the elimination and prevention of accidents in the electrical part of power systems [2] are, first of all, for express analysis.

Discusses software for disaster recovery of electricity supply in power systems and electrical networks. The implementation of these means is achieved through the use of expert systems formalisms and software considerations. It is shown that the deployment of recovery plans should occur at two levels: at the system level (operations with the equipment, in particular, from power lines) and patch level (operation switches). Examples of disaster recovery plans are considered.

2.2. OBJECT SYSTEM OF SUBSTATION ANALYSIS OF THE SITUATION

Scientific work in the field of operational dispatch analysis was started for a separate facility, 500 kV electrical substation, where Siemens systems are used as Automated Control System of Technological Processes (ACS TP). The "event sheet" formed in this process control system ACS TP is unnecessarily "long"-it can contain dozens of entries. The analysis of such a list is difficult for the duty personnel of the substation and is fraught with "human" mistakes. You can use special filters in the ACS TP to select only events that are important for a particular analysis. But then, before the analysis, the duty personnel must perform a special operation-the filter change is another possible source of errors in time deficit conditions. In addition, in the event sheet" of ACS TP not all the technological breakdowns necessary for the dispatch analysis (such as a breaker failure, failure of circuit breaker failure backup device, pull-in of a circuit breaker) are directly recorded.

To determine and describe the situation at the level of the power plant of an industrial enterprise, it is necessary to supplement the ACS TP of this power facility with a special expert system-substation analysis of the situation (SAS), and further develop these systems for the level of the electrical network (for example, facility of magistral electrical networks)-the SASEN system.

These expert systems are based on the MIMIR tool system developed in Russia-a small information model of intelligent solutions [3].

The system is named for the personage of Scandinavian mythology Mimir. With the wise head of Mimir, the supreme god Odin consulted in difficult situations.

In the MIMIR tool system, a database representation of semantic networks was adopted. There are two main sections of the database:

- topological model of the "primary" electric network,
- Logical model of relay protection and automation of relay protection and automation devices.

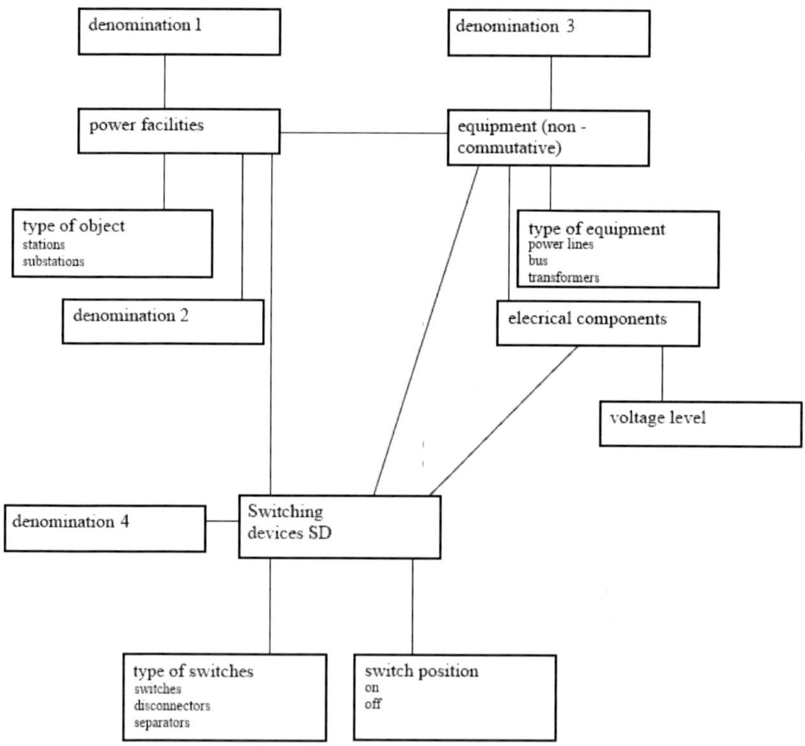

Figure 2.1. Topological model of the primary network.

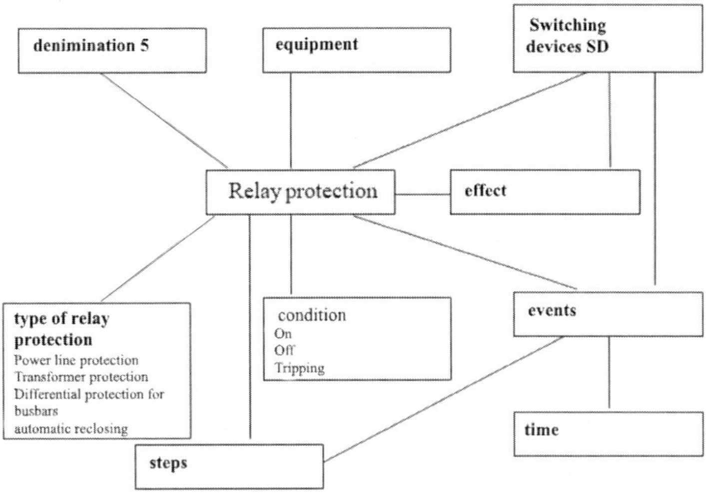

Figure 2.2. Logical model of relay protection.

The topological model in the generalized form can be presented in Figure 2.1, model RPA-in Figure 2.2.

The database is designed in such a way that access to it from the program can have the form of a question in a limited natural language with a dictionary of elementary concepts of the database.

For example, the question "Relay protection of Transformer № 3, event triggering?") gives as an answer an indication of the many protective transformers that have worked.

The database in MIMIR is supplemented by a system of programs in a limited natural language, simulating dispatching reasoning based on technological rules from [2].

The result is a language of reasoning programs, on which it is easy to set the logic of dispatching reasoning.

The rules on the basis of which the expert system of SASEN functions, are determined by the logic of the Relay protection. For example, if the protection of overhead transmission line is detected, the expert system determines which circuit breakers this protection is on. Have all the circuit breakers previously turned on from this set turned off. For non-switched off circuit breaker, the "circuit breaker failure" event is fixed and it is checked whether the circuit breaker backup device of this circuit breaker and the like has been activated.

In another, more complicated situation, when the transformer protection is triggered, it is necessary to determine whether there are defective switches and what type of defect (failure, tightening) takes place (unplugged circuit breakers must be disconnected before testing the extinguished buses by disassembling their circuit with disconnectors with lockout violation, and for the "tightened" switches, the circuit is disassembled after testing the tires). The tightening torque can be logically set in the argument-the breaker failure protection of the defective switch is triggered before it is turned off.

The results of the SAS operation can form the basis for the report of the on-duty substation sent to the network controller in the event of an abnormal situation. The experience of testing SAS at Chagino substation is positive.

To display the results of the expert system, the CASCADE-NT system is used [4].

2.3. Analysis System for Electrical Networks

The development of a system for the electrical network [5] allows to identify the damaged equipment elements, deactivated elements, de-energized parts of the network, failures in the operation of relay protections, failures and tightening of circuit breaker trips. The system is supplemented, in particular, by an analysis function that is quite complex, as is the situation of long-range redundancy [6]. With this analysis, the expert system first determines the "extinguished" buses of the substations and the disconnected lines, then the protection of the substations with the tires extinguished and on adjacent connections. The stages of the activated line protections are taken into account. At the substation, where there was damage, which caused the situation of long-range redundancy, the protection operates the first stage, and on the adjacent substations-the last steps. As an example [7], a fragment of an electrical network is considered after a short circuit on one of the lines and one of the circuit breakers (Figure 2.3).

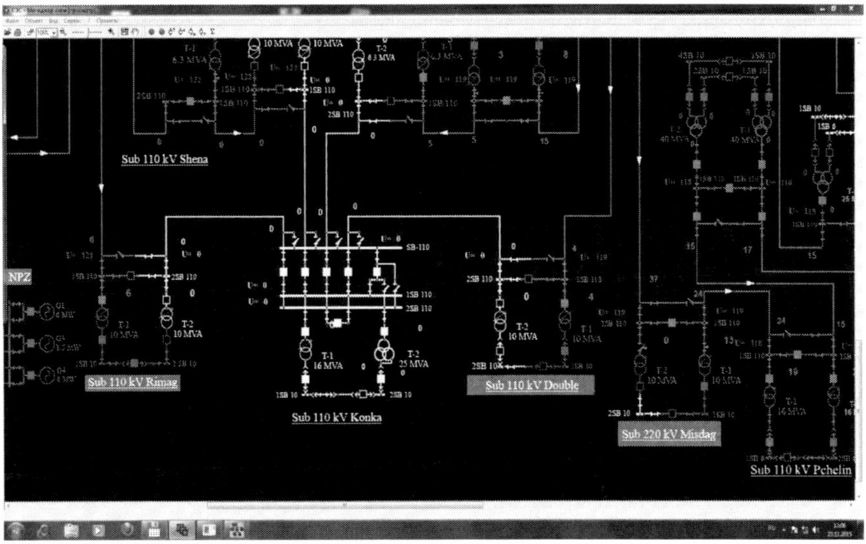

Figure 2.3. Example: a fragment of the circuit of the electrical network with a short circuit on the line VL 110 kV Konka-Lovinka. The white color indicates the sections of the network without voltage, where G1...Gn – generation center, Sub 110 kV – substation 110 kV, T1 16 MVA – transformer №1 16 MVA, 1SB – 1-st busbar system.

Here is a textual description of the situation synthesized by the SANS expert system.

--- Long-range backup control
Analysis of the situation of long-range redundancy
Situation: FURTHER RESERVATION
Presumably damage to equipment
VL-110kV Konka-Lolovinka
Reactor Relay Protection
Lovinka VL Konka stup. 1
Triggering line protection
RPL Konka VL 110kV Konka-Lovinka
Circuit breaker failure
Konka V 110 kV Lovinka
NO backup circuit breaker failure
Relay Protection Relay Operation at Neighboring Substations

For the "manual" analysis by the dispatcher this situation is very difficult: the protection was operated at five substations, the switchboards were disconnected at the substations, and there are no outages at the Konka substation, this substation was completely de-energized. The expert system of in this case (see the previous example) gives an exhaustive technological description of the situation, understandable to the dispatcher.

2.4. Intellectual Models of Reasoning in the Adviser of the Dispatcher of Electric Networks

With the help of special reasoning programs and on the basis of technological instructions [2], it is possible to form texts of advice on the actions of the dispatching personnel necessary in certain situations (and, first of all, in emergency and pre-emergency situations).

Example of formation of the council:

- Protection operation

Differential transformer protection (DPT) Konka T-1 16 MVA

- reserve transformer protection operation of the breaker failure protection after tripping of the transformer

Breakthrough Konka Konka V 110 T-1

- defective switch Konka V 110 T-1 breaker failure

--- TIP: turn off manually.

--- if the switch does not turn off, with the permission of the dispatcher

--- disconnect the line and bus disconnectors of this switch

--- with security lock violation

Konka

No backup transformer

The transformer is disabled by the backup protection

- Protection operation

Breakthrough Konka Konka V 110 T-1

- Possible circuit breaker failure

Konka V 110 T-1

Shutdown failure

At the substation Konka

The operation of the breaker failure protection is immediately preceded by the protection of the transformer

DPT Konka T-1 16 MVA

Konka T-1 16 MVA

Tire Redemption

Konka I SSh 110

--- NOT ALL connections switches connected to this bus

--- the switch did not turn off

Konka V 110 T-1

the manual disconnection of the circuit-breaker

--- if the switch is turned off, then with the permission of the dispatcher

--- try out-push to apply voltage to the redundant tires

- switch on the switch

Konka V-110 VL Revskaya

- upon successful testing of the school

Switch on the switches

Konka OV-110

Konka V110 VL Double

Konka T-1 10kV

TIP: Display repair for inspection and repair of damage

- transformer

Konka T-1 16 MVA

- and switch

Konka V 110 T-1

2.5. The Simulator of the Analysis of Non-Standard Situations

The use of the expert system as an on-line advisory system is hampered, for the time being, by the almost total absence in the UNEG of the operational transmission of information on the operation of protection from substations to dispatch centers.

At the same time, that an important stage of the dispatcher's activity is the analysis of the situation that has arisen as a result of technological disturbances, the work of relay protection and automation, automatic equipment shutdown-this stage is completely not covered by the existing training systems.

With regard to the activities of the dispatcher on the NCC and the UDR of electrical networks, the analysis of the situation should include at least the ability to identify damaged equipment elements (for example, short circuits on lines), deactivated equipment elements, de-energized sections of the network, failures in relay protection, failures and "tightening" of circuit breaker trips. This dispatcher should be able to make this analysis on the basis of available operational information: display of electrical circuits in the DEC with switch positions and parameter values received from telemechanics systems, reports of duty personnel of substations (tripping of relay protection devices, switching of switches). Situation analysis is carried out in the conditions of time deficit, and "human" mistakes in such analysis can be fraught with serious consequences. "Manual" preparation of appropriate training is laborious enough. Therefore, in order to check and maintain the appropriate qualification of the dispatching personnel, it is expedient to develop a special simulator for analyzing contingencies. Therefore, the development of the described direction was aimed at creating a simulator for analyzing contingencies. The lack or incompleteness of operational information on the operation of the protection is not an obstacle in the implementation of the simulator-simulation mode.

The use of the Advisor in the simulator version has an independent practical value-it opens the possibility of effective control and professional development of the dispatching personnel. Only after a comprehensive check in the simulator of functions, the adviser developed on the basis of MIIR can be included in the online work.

General requirements for the simulator of contingency analysis ANS.

The ANS simulator is designed for use in dispatching services of electrical networks and in personnel training centers.

The main functions of the simulator:

- computer support for the preparation of network emergency response drills (a convenient task of repair schemes and "test" damage);
- Automated check of dispatching personnel skills to analyze the situations associated with technological damage in electrical networks.

2.5.1. Simplicity of Preparation of Trainings

To train the trainer, the instructor should specify only the equipment to be repaired, equipment with technological damage (for example, a line with a short circuit), failed switches (indicating failure or pulling-out of the trip).

The naturalness of presenting the situation to the trainee. The switching state of the circuit is displayed on the operational diagram (as in the regime simulators), the lists of events are presented on-object-thereby simulating the receipt of reports from the duty personnel of the substations.

2.5.2. Automatism Analysis of the Situation

In the ANS, an automatic analysis of the situation specified in training preparation is used. In this case, the topological model of the network, the RPA model and the rules of logical inference are used, in the simulated situation during the training, possible technological violations (in particular, short circuits), switches that are turned off, disconnected items of equipment, separated areas of the network. The results of the analysis are automatically generated in the form of a protocol.

2.5.3. Convenience of the Workout

After the trainee reports on the completion of the instructor's training, the training system asks him a number of questions, for example, indicate (presumably) the damaged equipment item, set the protection function, indicate, if necessary, failures or "tightening" of the switches, indicate the operation of the recloser and breaker failure protection switching off the equipment and changing the position of the switches.

Based on the results of the trainer's answers, the *Training Description* of the situation should automatically be formed.

2.5.4. Objectivity of Evaluation of Training Results

The training system should automatically compare the "correct" description of the situation (the result of automatic analysis) with the training description. As a result of the comparison, an Evaluation Protocol should be formed containing definitions of the correct answers of the trainee, an indication of errors and the automatic issuance of penalty points for training.

2.5.4.1. Setting Training Conditions

When setting the training conditions for the ANS, the head of the training prescribes:

- repair mode of the network (repairs of tires, power transmission lines, transformers, switches);
- disconnection of circuit breakers from emergency control,
- a damaged component of the equipment (for example, a short-circuit power line),
- failures or "tightening" in the operation of the switches.

Repair jobs are made by selecting the menu of equipment, with automatically the necessary switching in the primary circuit and in the relay protection circuits (for example, when a single busbar system of substations is transferred to the repair with the transfer of connections and protections for the remaining bus system). When specifying repairs to the database of the expert system, a mechanism is used to automatically simulate repair orders.

2.5.4.2. Simulation of Reports from Substations

In the process of situation analysis, a list is formed-an event list containing texts of messages about the operation of relay protection and switching devices. To simulate messages from substation personnel in the process of network-wide training, this list is automatically “sorted out” into message lists from individual substations. These simulations of training reports can be called by the names of substations.

2.5.4.3. The Training Process

After getting acquainted with the training situation, the participants of the training are interviewed by the simulator to determine whether they recognize the situation correctly. Based on the participant’s answers, a *training description of the situation* is automatically generated, and the mistakes of the trainee detected by the expert system are automatically evaluated in penalty points.

2.6. Automation of Damage Search in Distribution Electric Networks

Possibilities for the development of expert systems technology in the field of operational dispatch analysis for more complex logical tasks show the solution of the task of constructing a dispatcher-advisor system for localization and elimination of violations in the operation of 0.38-20 kV distribution networks [9]. The technology for solving this class of problems is specified in the guidance material [10].

The dispatcher must take a decision on the procedure for eliminating the violation, taking into account the availability and location of personnel of grid enterprises, the possibility of attracting related organizations and consumers, the network scheme, the availability of transport and the possibility of travel, weather conditions, the degree of responsibility of disconnected consumers. The timeliness and accuracy of the decisions of operational personnel in the elimination of accidents depends on the efficiency and quality of the network company. Therefore, it is advisable to develop a software package for planning damage detection activities, including the management of the movements of the operational brigades. The task is multifactorial-along with the development of semantic structures for setting the network topology (specifying the presence of remote control of switches, location and indications of FIP fixing devices, location of the indicators of the damaged line section), a geoinformation part is required-the road graph with indication of their extent and connection with substations.

The resulting plan should contain actions for the analysis of electrical circuits, but provide for the implementation of these actions in such an order to minimize the number and "mileage" of movements of operational teams (depending on their initial location). In accordance with these conditions, special program-reasoning is developed. There is also a simulator mode when the trained user "manually" sets the plan for the localization of the damage, which in the control mode is compared with the reference plan compiled by the automated system.

2.7. Intelligent Functions of Automated Dispatching Management Systems

In a number of works [2-6], intellectual functions of automated dispatch control systems (ASDU) were considered. The effectiveness of these functions was shown in the formation of descriptions of situations and the issuance of advice for dispatch use personnel in abnormal situations. These functions use the platform of expert systems, and the formalism of the formation of *models of reasoning* by specialists in technology is applied [8].

The testing of the intelligent functions of the ASDU was carried out as part of the simulator of the dispatcher for supernumerary situations [3]. To realize the complex of ASDU operating in real time, it is necessary to consider the software architecture of this complex. At the same time, relying on a number of functionally different workplaces, it is advisable to distribute intellectual functions so that they correspond to the experience of various technologists and at the same time only a part of the overall task (in particular, the situation analysis) that was "adjacent" to the experience of the relevant specialists was performed. This approach corresponds to the concept of multi-agent systems of artificial intelligence [1].

2.8. Architecture of the Automated Dispatching Management System (ASDU)

The general functional structure of the multi-agent system is shown in Figure 2.4. The operational information from the control object (power system, electric network) is transferred to the Operational Information Complex (OIC) and is used in the Mode Model, supporting the display of the Mnemonic Chart. Information about significant events in the control object is automatically applied to the Mnemonic diagram with the help of special symbols. For example, the designations of disconnected equipment elements in case of short circuits in the network are shown in Figure 2.3. At the same time, information from the "Cascad ASDU" [7] is used.

More detailed analysis of events should be carried out by special intelligent programs and displayed in text form. For example, Figure 2.5 shows the results of the analysis of the situation in the electrical network.

Another example of situation analysis contains the texts of advice for the dispatcher (the fragment is given below):

- Protection operation DPT Konka T-1 16 MVA
- reserve transformer protection operation of the breaker failure protection after tripping of the transformer Konka Konka V 110 T-1

- defective switch Konka V 110 T-1 breaker failure
--- ADVICE: turn off manually.
--- if the switch does not turn off, with the permission of the dispatcher
--- disconnect the line and bus disconnectors of this switch
--- with security lock violation

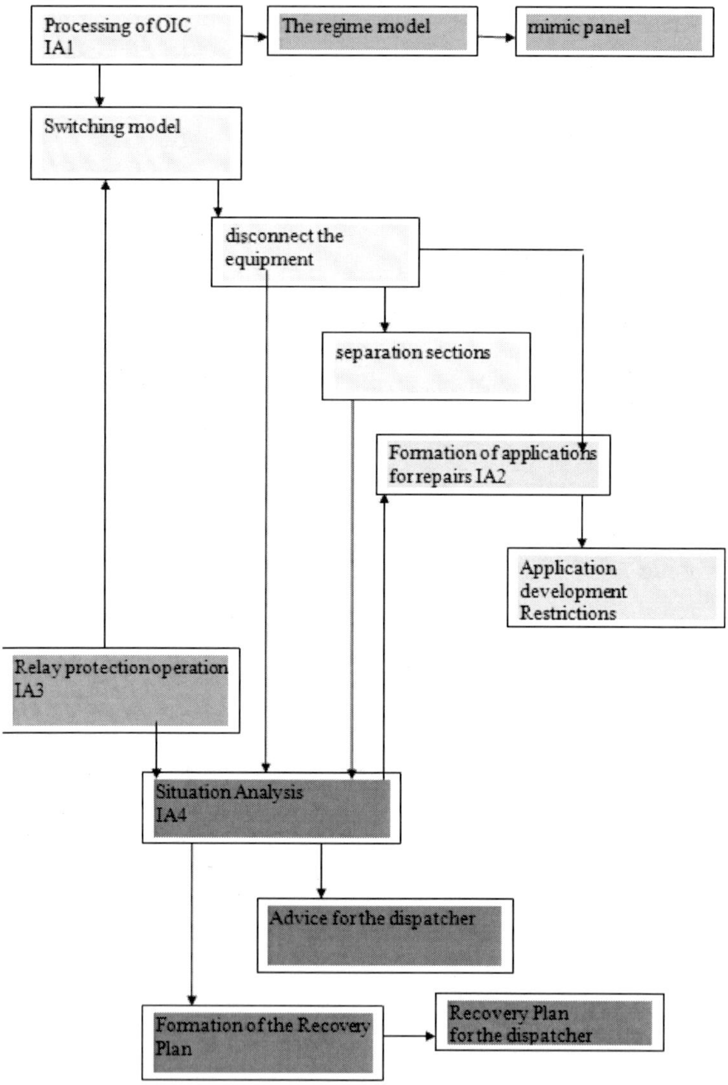

Figure 2.4. Functional structure of multi-agent ASDU.

2.9. INTELLIGENT AGENTS

Intelligent situation analysis programs for ASDU are functionally different and use different sections of system knowledge. It is advisable to organize the execution of these programs by different intellectual agents-relatively simple modules, each of which uses its own knowledge area and has its own initiation conditions. Since ASDU, as a rule, is realized in the form of a computer network, intelligent agents are localized in computerized workplaces for various specialists [9].

The structure of Figure 2.4 contains at least 4 intelligent agents:IA1 is a cyclically functioning agent. Based on the results of the operational processing of information in the OIC (switch positions), it forms the current switching model of the network, records the equipment shutdown events and the separation of network sections. Uses information about the operation of relay protection devices from IA3;

For the equipment shutdown events and the separation of network sections, voltage nodes (nodes corresponding to the generators of some stations, as well as connections to the integrated power system or neighboring power systems) in IA1 are automatically assigned. In the IA1 module, the connections of the nodes of the specified equipment to the source nodes are checked through switched switching devices. If there are nodes that are not connected to the sources, then the connections of these nodes through the switching devices determine the separated parts of the network.

IA2-agent "Repairs," initiated by fixed in IA1, as well as events that are processed in IA4 about disconnection of equipment, considers (emergency) requests for disconnection of relevant equipment elements, correcting relevant regime restrictions if necessary [6];

When automatic online review of repair orders, the instructions of the regime stored in the knowledge base are used. The instructions set limits on the mode parameters (usually the active power in the sections), which must be observed when the equipment specified in the application is disconnected. These limitations can be of two types: imposed for the time of application and imposed for switching. The discrepancy between the restrictions on the application being processed and the restrictions on the applications that are already allowed can lead to the abandonment of one of the conflicting applications or to the establishment of the most serious restriction on the "overlapping" of applications. At the termination of repair under already approved order because of failure the time necessary for change of schemes of repair of corresponding objects is considered. Module IA2 provides controllers with a table of repairs with appropriate restrictions.

IA3-an agent of "Relay Protection and Automatic Equipment" (RPA), functions cyclically, revealing the events of operation of relay protection devices and object automatics;

In the IA3 function, in addition to other specific tasks, the formation of events related to the operation of relay protection devices and PA emergency control devices is included. The source of this information is the object data concentrators. Since this article deals only with the dispatcher's quick analysis of situations, relay protection and PA events in which it is important not to accurately record the event time, but the "correct" sequence of events, the events of the relay protection devices and PA have the same status as the switching devices, in IA1. In the future, with the automation of full analysis, IA3 functions can be more complex.

IA4-the "Analysis" agent is initiated when the RPA tripping, equipment shutdowns, and sections of the network sections are committed. This agent generates a description of the situation and (if necessary) displays the texts for the dispatcher. Principles of organizing a logical dispatching rapid analysis of the analysis of situations are described in the literature [2-5], [9]. Let us briefly describe the functions IA4, not described in these papers.

For abnormal situations, the Plan for Restoring Normal Network is formed in IA4.

2.10. Specialization of Intellectual Agents

When drawing up this plan, objects (substations) needing restoration are identified, as well as objects-power supply centers (PSC). For each of the objects to be restored, the "recommended" PSC and the minimum possible recovery path are determined (by the principle of "electrical proximity to the PSC"). The plan should include the switches of normally disconnected switches belonging to the trajectory. For circuit breakers disconnected by an open repair request, the plan provides for the "curtailment" of repair work for this application through the emergency response time, and (for an unacceptably long time), the presence of detour paths for the path section containing the repairable switch is considered. For the prepared plan, the number of necessary commutations and the maximum emergency response time are calculated if it is necessary to "fold" the repairs.

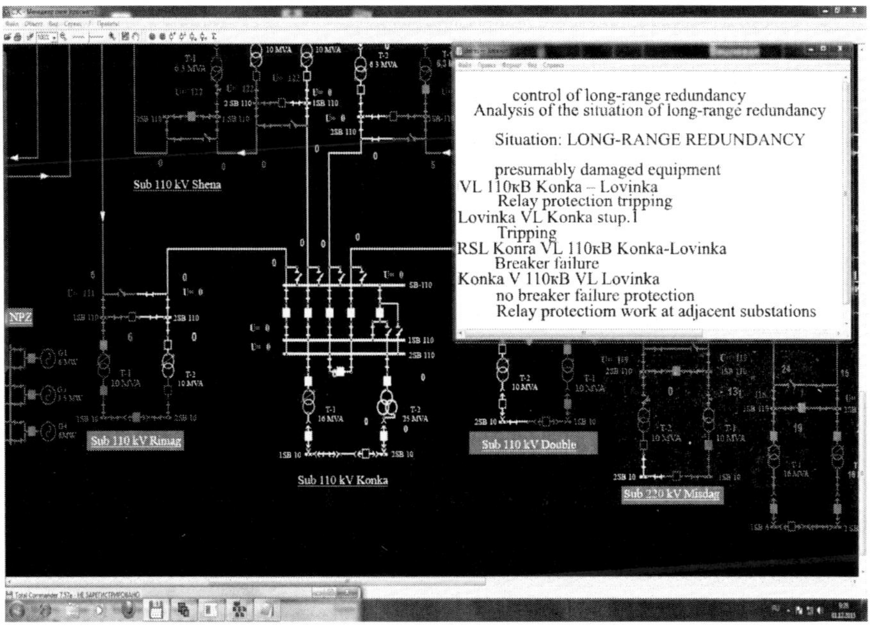

Figure 2.5. Text display of the results of the situation analysis.

The special module included in the IA4 determines whether the situation of "long-range backup" is: when unilateral operation of relay protections of lines, adjacent objects are checked to identify among them an object on which there were no trips of protection).

The presence of a long-range backup situation must be recorded in the text of the situation analysis protocol.

Localization of software modules of intelligent agents:

IA1-the workplace of the Operational-Information Complex (OIC),
IA 2-a workplace of the expert on electric modes,
IA3-the workplace of the "relay operator"
IA4- workplace dispatcher.

Each of the intellectual agents must form a List of Events within its competence.

Figure 2.5 shows the results of the analysis of the situation in the electrical network.

2.11. Automation of After-Emergency Control

Automation of dispatching management of power systems should contain means for automation of emergency control, as one of the most important tasks. The relevance of this task is confirmed by the cases of complex cascading accidents in domestic power systems.

In the general case, an "archipelago" is formed from the "islands"-the districts that have separated from the Unified Energy System (UES) after the accident. It is necessary to define a recovery plan for the maximum possible satisfaction of customer requirements while minimizing losses.

At present, there are not practically universal software tools for forming a recovery plan. Therefore, in emergency situations, the recovery is practically done manually by the dispatch personnel. In this case, "human" errors can lead to serious consequences. The use of intelligent systems (expert systems, program-reasoning) will allow the automatic formation of such plans. Instead of traditional software methods, such systems use a system of rules formulated in the natural language of a technologist. It is expedient to create a recovery plan in the intelligent system of the

Automatic Dispatch Control System (ADCS) on two levels:

1) System level-action plan with UES equipment

2) Switching level-a plan for commutation switching elements.

2.12. System Level in Planning the Restoration of Electric Power Supply

The literature [1, 11] notes the complexity of the problem of recovery planning, the absence of a universal algorithm. Implementation based on the use of expert systems and reasoning programs [5, 12] makes it relatively easy to build a system for planning this process. It is significant that in this case, thanks to the application of heuristic rules, it is possible to provide a simple modification of the system, if additional parameters and operations are to be taken into account. The general algorithm for developing a recovery plan is shown in Figure 2.6. The algorithm is simplified-in particular, there is no indication of operations with generators in redundant areas.

The "Recovery" program that implements this algorithm is part of the SmartCom subsystem of the "Dispatcher advisor" complex. The program is designed for:

- identification of areas separated from the main network,
- the formation of a description of the separated areas,
- formation of the Emergency Repair Plan.

The knowledge structure for the task contains the concepts needed to describe the energy system (objects, types of facilities, equipment, types of equipment, switching devices, types and positions of switching devices, electrical network nodes, voltage levels [5]). To solve the reconstruction problem, this structure is complemented by the semantic structure shown in Figure 2.7.

The questions are used to identify the connections of the equipment elements through electrical nodes and switching devices to identify the separated areas and their links. The operation of the algorithm is illustrated by the example of a fragment of the 500 kV network diagram of Central region (Figure 2.7). The scheme is used when conditionally specifying parameters-generations and loads. In Figure 2.8 power objects of the separated region are highlighted in color (the First substation scheme in the example is divided due to the disconnection of the bus-couplers). Disconnected power lines are shown as dotted lines.

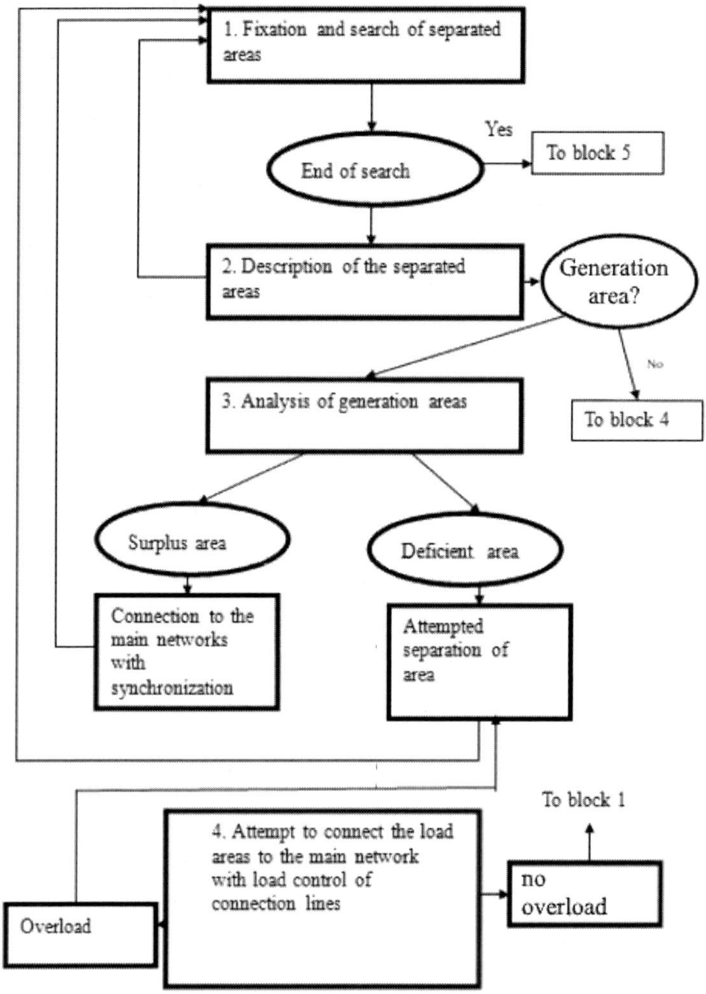

Figure 2.6. General recovery of power supply algorithm.

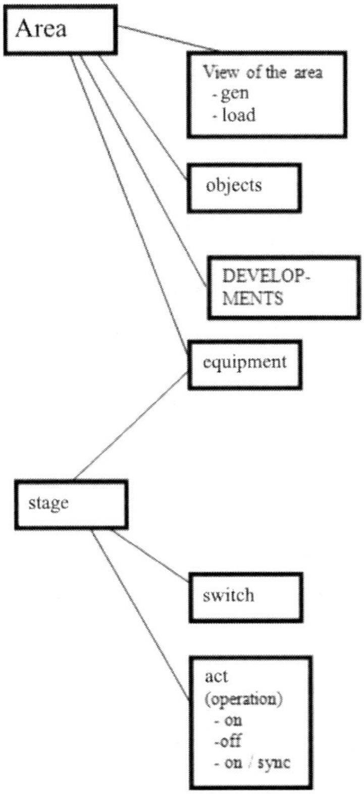

Figure 2.7. Additional semantic structure for the task of recovery of power supply.

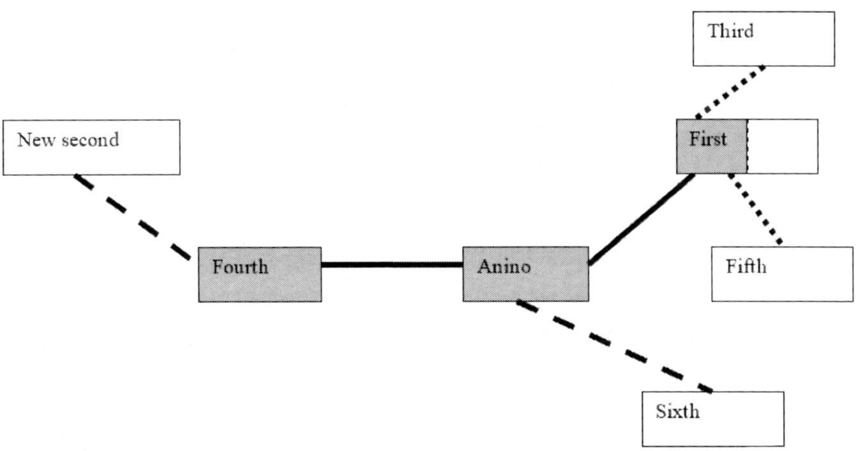

Figure 2.8. Branch of the district in the network structure of 500 kV of Central region.

An example of a system operation protocol for forming a plan according to the algorithm is here in Figure 2.6. When considering the protocol, it should be borne in mind that it describes a multi-stage process and the identified separated regions are renumbered at each stage [12-13]. UES-Unified Energy System.

Simplified protocol of the algorithm:

SEPARATION FROM UES
--- Area 1 ---
First-the scheme of the object is divided with the UES
Anino
Fourth
--- Description of the area 1 ---
First-the scheme of the object is divided with the UES
Anino
Anino
Fourth
--Communication of the region with the UES through the switches First
- Lines for communication of the area with the UES
L 500 Sixth – Anino
VL 500 New second– Fourth

With the accession of Area 1 to the UES
Overload of the line is possible
VL 500 New second– Fourth
[Comment: the system tries to attach the separated region to the UES. Possible
overload of the line is identified. Then the system tries to divide the separated
region and attach its parts to the UES alternately]
Attempt to divide the area 1
----SWITCH OFF
VL 500 First – Anino
----SWITCH OFF

VL 500 Anino- Fourth

SEPARATION FROM UES

--- Area 1 –

Fourth

--- Area 2 ---

First-the scheme of the object is divided with the UES

--- Area 3 ---

Anino

--- Description of the area 1 ---

Fourth

- Lines for communication of the area with the UES

VL 500 New second– Fourth

- Lines for communication of Region 1 with Area 3

VL 500 Anino- Fourth

- No stations in the area 1

Fourth

--- Description of the area 2 ---

First

- Lines for communication of he area with the UES

VL 500 First – Third

VL 500 Fifth West with the cons.

VL 500 First – Anino

- Lines for communication of Region 2 with Area 3

VL 500 First – Anino

- No stations in the area 2

First

--- Description of the area 3 ---

Anino

- Lines for communication of the area with the UES

VL 500 First – Anino

VL 500 Sixth– Anino

--Links for communication of Region 3 with Area 1

VL 500 Anino- Fourth

--Links for communication of Region 3 with Area 2

VL 500 First – Anino

- No stations in the area 3

To join Area 1 to the UES

--CLOAD

VL 500 New second– Fourth

For the accession of Area 2 to the UES

--CLOAD

VL 500 Fifth Zap. with the cons.

For the accession of Area 3 to the UES

--CLOAD

VL 500 Sixth– Anino

No disconnections from the UES

[Comment: the final recovery plan generated by the intelligent system is displayed]

REHABILITATION PLAN:

--- 1 . DISCONNECT

VL 500 First – Anino

--- 2. DISCONNECT

VL 500 Anino- Fourth

--- 3. ENABLE

VL 500 New second– Fourth

--- 4 . ENABLE

VL 500 Fifth Zap. with the cons.

--- 5 . ENABLE

VL 500 Sixth – Anino

--- 6. ENABLE

VL 500 First-Anino

--- 7. ENABLE

VL 500 Anino- Fourth

The received recovery plan should be (step-by-step) checked on the regime model (for example, [4]) to identify possible unacceptable values of the parameters (in particular, for detecting inadmissible voltage deviations in electric nodes).

In this (as much as possible simplified) version of recovery planning does not provide for the optimization of plans, does not take into account the availability of system automation-these are issues that will be resolved with further development of the concept outlined here.

2.13. Switching Level

The automatic development of the switchgear recovery plan is considered using the example of a fragment of the power grid of the energy system of South region (Figure 2.9) in case of a simple accident (short circuit on the overhead transmission line with the failure of the linear circuit breaker and the operation of circuit breaker failure backup device).

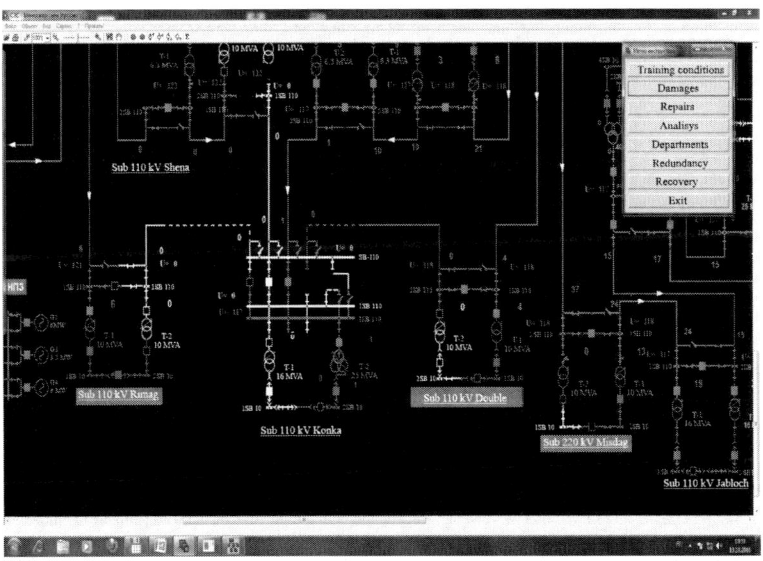

Figure 2.9. The repair of 110kV Konka-Rimag, then the short-circuit at 110kV Konka-Revskaya, with the failure of a line breaker at Konka.

It is necessary to develop a plan for the restoration of the electric network-a sequence of switching operations for transferring the network to a state in which the maximum power supply of consumers is ensured. It is intuitively clear that this finite restored state should be as close as possible to the initial (pre-emergency) state of the network [12, 13]. If these conditions are met, the best recovery plan should contain the minimum number of transactions. In this case, operations are not only switching, but also the initiation of applications for the repair of equipment elements [14-19].

The recovery algorithm contains:

1) Definition of the situation:
2) Damage to equipment (SC)
3) Failure and tightening of switches
4) Repayments at substations
5) Definition of recoverable objects
6) Definition of CPUs (network power centers)
7) Heuristic definition of the recommended CPU for recovery,
8) Heuristic definition of the order of restoration (on objects)
9) Initiation of an application for the repair of damaged equipment
10) Initiating an application for repair of a failed switch
11) If the breaker failure device fails, "check the condition of the breaker failure protection"
12) Restoring objects
    12.1) Disabling the damaged line LR
    12.2) Switching on of the circuit-breakers disconnected in case of an accident
    12.3) Finding a trajectory (equipment) from the CPU to the object
    12.4) Checking the position of the switches along the path
    12.5) Turn on the disabled circuit breakers along the trajectory (taking into account the applications)
    12.6) For switches according to clause 9.5, for which there are orders, ask "to cancel the work on the application within the AG-emergency preparedness period," determine (if possible)

"workarounds"-for an unacceptably long AG time for an application for an on-off switch.

13) Counting of operations, orders, AG- emergency preparedness [20-38].

=== PLAN OF RECOVERY OF THE DETACHED NETWORK ===

A short circuit is on the equipment

VL-110kV Konka- Revskaya

Circuit breaker failure

Konka V-110 VL Revskaya

breaker failure protection

Repayments on substations:

Lovinka

Konka

Double

Revskaya

Rimag

Recommended nutrition center for restoration: Mysdag

1. Initiate a request the repair of equipment

VL-110kV Konka-Revskaya

2. Try to disable the key

Konka V-110 VL Revskaya

If it does not turn off, disconnect it with disconnectors with violation of the lock

3. Initiate a request to repair a circuit breaker

Konka V-110 VL Revskaya

4. Restore an object Rrevskaya

==== PLAN OF RECOVERY FOR SUBSTATION Revskaya

4. 1. Disable LR

Revskaya LR-110kV VL Konka

4.2. Switch on the switch

Revskaya B-110 T-1

4.3. Switch on the switch

Revskaya SV-110

==== Recovery from the center ====

Mysdag

5. Restore an object

Konka

==== PLAN OF RECOVERY FOR SUBSTATION

5.1. Disable LR

VL-110kV Konka-Revskaya

5.2. Switch on the switch

Busbar connected circuit breaker 10 кВ

5.3. Switch on the switch

Konka V110 VL Double

5.4. Switch on the switch

Konka Busbar connected circuit breaker 110kV

5.5. Switch on the switch

Konka V 110 T-1

==== Recovery from the Mysdag center

Comment: The program determines the trajectory from the power center to the renewed substation and checks the position of the switches along the path; for disconnected circuit breakers, the presence of repair orders is checked

VL-110kV "Konka- Double"

HVL 110 kV "Mysdag-Double"

110 kV overhead line "Mysadag- Double" slot traction

Mysdag-"Double"

---- there must be a link to the line

110kV Konka-Anchor slot through the substation

"Double"

"Double" CB-110

---- switch on

Mysdag VE VL 110 kV "Double"

---- switch on

"Double" of RP-110

---- switch disconnected

VL110 kV “Konka- Double”
Konka
“Double”
Konka V110 VL Double
---- switch disconnected
“Double” of RP-110
No repairs
5. 6. Switch on the switch
“Double” of RP-110
Konka V110 VL “Double”
No repairs

Commutations 10
Orders 1
AG Sum.

CONCLUSION

1) It is advisable to include in the ADCS substations and electrical networks of industrial enterprises of intelligent subsystems using the technology of expert systems. These subsystems, whose main functions are considered in this article, contain semantic structures for describing the topology of control objects and special program-reasoning based on technological instructions. Intelligent subsystems provide computer support for operational personnel in the analysis of contingencies in control objects. In addition to the functions of the system-advisors, an important new function is provided-the construction of dispatching simulators for analyzing contingency situations that ensure the verification and maintenance of the necessary qualifications of the dispatching personnel of electrical networks. In this case, the simulator function is provided even if there is a deficit or lack of operational information about the operation of the relay protection devices.

2) It is expedient to use the multi-agent structure of intellectual functions during constructing the ASDU complexes by power systems and power networks.
3) Intellectual agents should be localized in the workplaces of technologists.
4) The automation of post-emergency recovery in power systems and in electrical networks is advisable with the help of intelligent systems in the practical absence of traditional universal programs. The use of methods of expert systems and reasoning programs in these systems makes it possible to create flexible recovery algorithms based on heuristic rules on dispatching experience.
5) Recovery algorithms should provide the ability to deploy recovery plans at two hierarchical levels: system (actions with equipment such as power lines) and on switching (actions with switches).

ACKNOWLEDGMENTS

To Dr. Yuriy Ya. Lyubarsky, Russia, R&D Center of FGC UES, Mr. Nikolaj M. Alexandrov, post-graduate student, Russia, JSC "Dinamika," Cooperation of Universities and Innovation Development, Doctoral School project "Complex diagnostic modeling of technical parameters of power transformer-reactor electrical equipment condition," 2009 has made publishing of this article possible.

REFERENCES

[1] Anashkin, S. V., Kartashev, S. V., Lyubarsky, Yu. Ya., Miroshkin, A. G. Automated dispatching analysis of abnormal situations in electrical networks. *Electrical stations*, 2013, № 9. p. 49 (in Russian).

[2] National Standard. *Instruction for the prevention and elimination of accidents in the electrical part of power systems.* Moscow: Published in ENAS, 2004(in Russian).

[3] Lyubarsky, Yu. Ya. *Intellectual information systems.* Moscow: Nauka, 1990 (in Russian).

[4] Rabinovich, M. A. *Display of operational information. Complex CASCADE-HT.2.0.* Moscow-Publishing House ENAS, 2004 (in Russian).

[5] Anashkin, S. V., Kakovskiy, S. K., Kartashev, S. V., Lyubarsky, Yu. Ya., Miroshkin A.G. The simulator for the dispatching personnel of electric networks. *Electrical stations*, 2014, № 2 c. 2-6 (in Russian).

[6] Taubes, I. R. *Breaker failure backup device.* Moscow: Energoatomizdat, 1988 (in Russian).

[7] Kakovskiy, S. K., Lyubarsky, Yu. Ya. The software system is the adviser for managing electrical networks. *Electrical stations*. 2016 № 2 p. 35-39 (in Russian).

[8] Lyubarsky, Yu. Ya. Intelligent modes of reasoning in the advisor of the electrical network manager. *Electrical stations.* 2017 № 3 p. 35-39 (in Russian).

[9] Vlasyuk, V. D., Kakovskiy, S. K., Lyubarsky, Yu. Ya. Automation of damage search in distribution electric networks. *Electrical stations 2015* № 4 p. 29-36 (in Russian).

[10] National Standard. Typical instruction on elimination of violations in the operation of distribution networks 0.38-20 kV with overhead power lines. *Typical instruction* 34-70-058-86 (in Russian).

[11] Gurevich, L. A., Vakhitov, A. N. Multiagent systems. *Doing in Computer Science*-2005 p. 116-139 (in Russian).

[12] Lyubarsky, Yu. Ya. Operative dispatching analysis of abnormal situations in electrical networks of industrial enterprises. Use of computer support on the basis of technology of expert systems. *Industrial Energy*, 2017, No. 9 (in Russian).

[13] Kakovskiy, S. K., Lyubarsky, Yu. Ya. Software system-adviser for the management of electrical networks. *Electrical stations*, 2016, № 2 (in Russian).

[14] Lyubarsky, Yu. Ya., Portnoy, M. G., Rabinovich, R. S., Leviush, M. A., Kniazeva N. P. Expert system for the operational review of repair orders for the ASDU of energy associations. *Electricity,* 1991, № 2 (in Russian).

[15] Lyubarsky, Yu. Ya., Khrennikov, A. Yu., Aleksandrov, N. M. Unplanned situations in electrical networks. Operational dispatching analysis based on technology of expert systems. *News of Electrical Engineering,* №. 4 (106), 2017, p. 18-21 (in Russian).

[16] Thanh, B. D., Voropai, N. I. Restoring the power supply system with distributed generation after a major accident. *Industrial Energy,* 2011- No. 8, pp. 12-18 (in Russian).

[17] Uspensky, M. I., Kyzrovev, I. V. *Methods of restoration of power supply in distribution networks.*-Syktyvkar, 2010m 124 p. (Komi Science Center, Ural Branch of the Russian Academy of Sciences) (in Russian).

[18] CIGRE SC 8 WG 38.08 TF04: *The Use of Expert System for Power System Restoration* (Final Report (Draft) Feb 1994.

[19] *IEEE Transactions on Power Systems*, Vol. 11, No. 1, February 1996, 383-389. Knowledge-Based Behavior Interface: Application to Power Network Restoration Support System.

[20] Zhang, J. J., Wang, F.-Y., Fellow, A., Wang, Q., Hao, D., Yang, X., Gao, D. W., Zhao, X., Zhang, Y. Parallel Dispatch: A New Paradigm of Electrical Power System Dispatch. *Journal of Automatica SINICA*, vol. 5, NO. 1, January 2018.

[21] Vadari, S. *Electric system operations: evolving to the modern grid.* Artech House, 2012. http://www.geocities.ws/smartcitiesmauritius/Vadari_flyer_2015.pdf.

[22] Caramanis, M. C. "*It is time for power market reform to allow for retail customer participation and distribution network marginal pricing,*" http://resourcecenter.smartgrid.ieee.org/sg/product/publications/SGNL0060.

[23] Farhangi, H. The path of the smart grid. *IEEE Power Energy Mag.*, vol. 8, no. 1, pp. 18-28, Jan. -Feb. 2010. http://ieeexplore.ieee.org/xpls/icp.jsp?arnumber=5357331.

[24] Brown, L. R., Adams, E., Larsen, J., and Roney, J. M. *The Great Transition: Shifting from Fossil Fuels to Solar and Wind Energy*. New York: W. W. Norton & Company, 2015. http://ci.nii.ac.jp/ncid/BB18640670.

[25] Ackermann, T., Andersson, G., and Söder, L. Distributed generation: A definition. *Electr. Pow. Syst. Res.*, vol. 57, no. 3, pp. 195-204, Apr. 2001. http://www.sciencedirect.com/science/article/pii/S0378779601001018.

[26] Vardakas, J. S., Zorba, N., Verikoukis, C. V. A survey on demand response programs in smart grids: Pricing methods and optimization algorithms. *IEEE Commun. Surv. Tut.*, vol. 17, no. 1, pp. 152–178, 2015. DOI:10.1109/COMST.2014.2341586.

[27] Cheng, X., Zhang, R. and Yang, L. Q. Consumer-centered energy system for electric vehicles and the smart grid. *IEEE Intell. Syst.*, vol. 31, no. 3, pp. 97-101, May. -Jun. 2016. http://ieeexplore.ieee.org/document/7478506/.

[28] Cheng, X., Hu, X. Y., Yang, L. Q., Husain, I., Inoue, K., Krein, P., Lefevre, R., Li, Y. Y., Nishi, H., Taiber, J. G., Wang, F. Y., Zha, Y. B., Gao, W., and Li, Z. X. Electrified vehicles and the smart grid: The its perspective. *IEEE Trans. Intell. Trans. Syst.*, vol. 15, no. 4, pp. 1388-1404, Aug. 2014. http://ieeexplore.ieee.org/xpls/abs_all.jsp?arnumber=6869118.

[29] Zhang, Y. J. *Automation of Power Grid Monitoring and Dispatch*. Beijing: China Electricity Press, 2007.

[30] Wang, S. Z. *Power Grid Dispatch and Operation Techniques*. Shenyang: Northeast University Press, 1997.

[31] Fish, A. N. *Knowledge Automation: How to Implement Decision Management in Business Processes*. Hoboken, NJ USA: Wiley, 2012.

[32] Barker, P. P. and De Mello, R. W. "Determining the impact of distributed generation on power systems. I. Radial distribution systems, In Proc. *IEEE Power Engineering Society Summer Meeting*, Seattle, WA, USA, vol. 3, pp. 1645-1656, 2000. http://ieeexplore.ieee.org/document/868775/.

[33] Deng, J. L., Wang, F. Y., Chen, Y. B., and Zhao, X. Y., "From industries 4. 0 to energy 5.0: Concept and framework of intelligent energy system," *Acta Autom. Sin.,* vol. 41, no. 12, pp. 2003-2016, Dec. 2015.

[34] Wang, F. Y. "Parallel control: A method for data-driven and computational control, *Acta Automat. Sin.*, vol. 39, no. 4, pp. 293-302, Apr. 2013. http://en.cnki.com.cn/Article_en/CJFDTOTAL-MOTO20 1304002.htm.

[35] Wang, F. Y. and Wong, P. K., "Intelligent systems and technology for integrative and predictive medicine: An ACP approach," ACM *Trans. Intell. Syst. Technol.* (TIST), vol. 4, no. 2, pp. Article No. 32, Mar. 2013. http://www.ncbi.nlm.nih.gov/pmc/articles/PMC4039638.

[36] Goodfellow, I., Bengio, Y., Courville, A., *Deep Learning*. Cambridge, UK: The MIT Press, 2016.

[37] Wang, F. Y., Zhang, H. G., Liu, D. R. Adaptive dynamic programming: An introduction. *IEEE Comput. Intell. Mag.*, vol. 4, no. 2, pp. 39–47, 2009. DOI:10.1109/MCI.2009.932261.

[38] Sato, T., Kammen, D. M., Duan, B., Macuha, M., Zhou, Z. Y., Wu, J., Tariq, M., and Asfaw, S. A., *Smart Grid Standards: Specifications, Requirements, and Technologies.* London, UK: Wiley, 2015.

Chapter 3

PARTIAL DISCHARGES AS DIAGNOSTIC METHOD FOR DETECTION OF ELECTRICAL EQUIPMENT AND XLPE CABLE FAULTS, DEFECTS FOR POWER SYSTEMS CONTROL

A. Yu. Khrennikov*, PhD

ABSTRACT

This chapter presents an application's experience of monitoring of the electric activity of the partial discharges (PD) into the insulation of transformers, data of electrical tests. The determination of partial discharges intensity was produced with the aid of the electrical method of PD registration. As the measuring means was used digital oscillograph, and also digital registrar-analyzer of the partial discharges. As a result the internal inspection of autotransformer windings after emergency switching off was revealed the damage and short-circuit of the forth turns of regulating winding of 167 MVA/500/220 kV autotransformer. Methods and equipment for testing XLPE cables by voltage on the ULF (0.1-0.01 Hz.) are also considered in the article.

* Corresponding Author's E-mail: ak2390@inbox.ru.

Keywords: partial discharges (PD), digital registrar-analyzer, electrical method of PD registration, internal inspection, autotransformer, XLPE cables, ULF for testing

3.1. Introduction

A large part of the electrical equipment of electrical stations, substations and power transmission and distribution systems has developed a service resource, but continues to operate, because its replacement requires more funding. Therefore, every year the increasing cost to undertake comprehensive surveys and diagnostics. Assessment of the actual status of power electric equipment based on diagnostic measurements is very difficult and pressing challenge.

Infra-red (thermal) control is one of the main directions of development of the system of technical diagnostics, which provides the ability to control the thermal state of electrical equipment and substation's buildings without taking them out of work, identify defects at an early stage of development, reducing the cost of technical surveys by predicting the timing and extent of the repairs [1, 2, 3].

Electrical equipment, in particular power transformers, is one of the basic parts in the circuitry of power transmission and delivery. Therefore the interest to perfection of electrical equipments' fault diagnostic methods is being increased. The repairs of power transformers and other electrical equipment are carried on, using diagnostic measurement results [1, 2, 3, 5-9].

Also this chapter presents inspection of insulation current transformers by infra-red (thermal-vision) control, measuring the activity of the partial discharges (PD) into the insulation of transformers, data of electrical tests, the unreliable fastening of protective hoods on the diaphragm devices, the defects of installation.

3.2. Partial Discharges (PD) for the Insulation Monitoring

Breakdown of the turn-to-turn insulation of the power transformers as a result of the action of partial discharges (PD)

The internal short-circuits of transformer windings can be caused by the breakdown of the turn-to-turn insulation as a result of the destruction of insulation under the action of operational factors and by the action of the partial discharges in the place of future breakdown, as initiator of which can serve commutation, thunderstorm and other increased actions on the transformer insulation. The tracks of partial discharges during the insulating construction are represented in Figure 3.1.

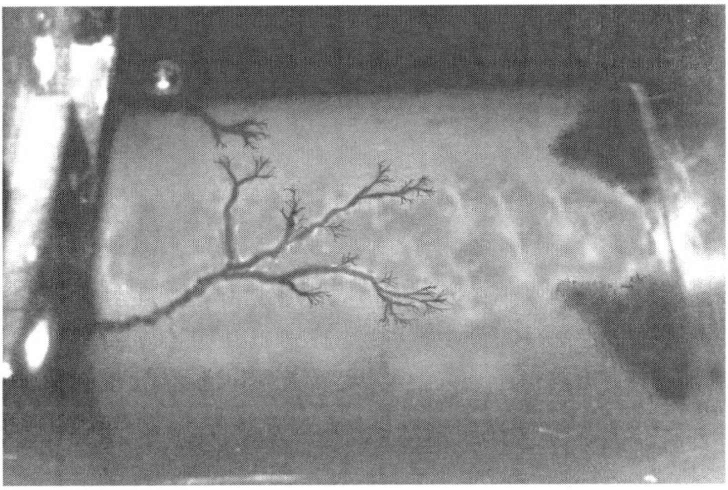

Figure 3.1. The tracks of partial discharges during the insulating construction.

All results of the traditional preventive insulation routine tests of autotransformer according to National testing Standard, the preceded moment of damages, gave conclusion about satisfactory condition state of insulation.

The specialists of the Novosibirsk specialized testing laboratory carried out the unplanned measurements of PD intensity on the autotransformer.

Although the PD measurements on the autotransformer were carried out not fully, only on the 220 kV winding side as a result of turning off of sensors along the side 500 kV, that it did not make it possible to accurately localize defect, also in this case were discovered discharge phenomena, were given recommendations about conducting of revision of tap-changer. Revision was not carried out.

Preliminarily possible reasons of the damages of the regulating winding of autotransformer by way of the decrease of probability could be:

1. Malfunctions in the tap-changer, which led to the appearance of the increased currents in the regulating winding, and, as a result, to local overheating and to the destruction of turn winding insulation.
2. Winding deformation with the decrease of turn-to-turn distances as a result of close short-circuits, occurred after previous repair season.
3. Breakdown of turn-to-turn insulation as a result of the lightning surge.
4. Manufacturer defect in the region of one or two turns (agnail, the spikes of winding wire).
5. The entry of foreign objects into the region, where turn-to-turn short-circuit occurred.

The results of the measurements of PD intensity on the autotransformer before the damage are given below [by 15-18].

3.3. Procedure of PD Measurements and Measuring Means

The determination of PD intensity was produced with the aid of the electrical method of PD registration. As the measuring means was used digital oscillograph TDS-2204 "Tektronix", and also the eight-channel digital registrar-analyzer of the partial discharges "R2000" of production

"Vibro-center", Perm' city. The diagram of PD registration in each phase AT is given to Figure 3.2.

It was used as the sensors the stationary devices of connection of the type UP-6. It was impossible to use 500 kV sensors, since continued the process of the agreement of the joint operation of the device of the connection UP-6 with KIV-500, and 500 kV sensors were switching off.

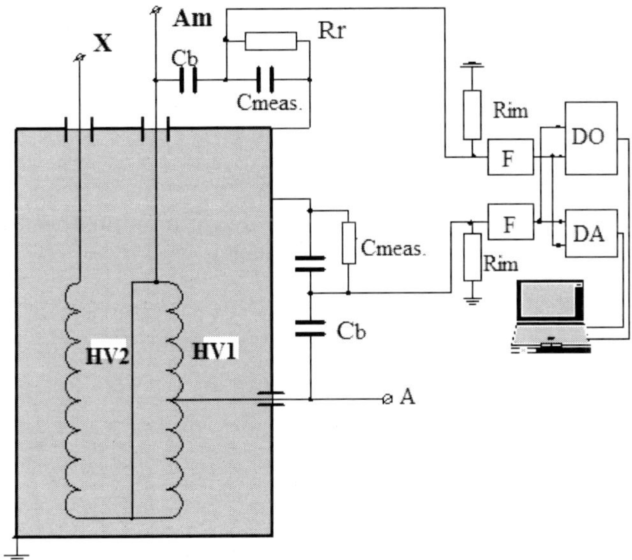

Figure 3.2. Diagram of PD registration of 167 MVA/500/220 kV autotransformer: Rr – measuring resistance, 50 Ohm; Rim – matched impedance, 50 Ohm; F – the filter of high frequency; Cb – the capacity of 500 and 220 kV bushings; Cmeas. – the capacity of the measuring facing of 500 and 220 kV bushings; DO – digital oscillograph; DA – digital analyzer.

The procedure of PD intensity measurement with the aid of the digital oscillograph (DO) consisted of the following: the measured signals from the sensors of different phases were investigated. It was produced the filtration of signals by series connection into the chain of high-pass filters in this case, the agreement of the wave drag of cable with the load, changes in polarity and level of the starting of development, scales along the axes of oscillograms and etc.

PD selection was produced on the specific criteria, which it is discussed below. With the use of a digital analyzer was produced the more thorough selection of PD pulses from the noise signals. Recorded in the digital signal aspect underwent the subsequent the processing by special software for determining the PD fundamental characteristics.

3.4. Results of Partial Discharge Measurements

The signals from the 220 kV sensors during several periods 50 Hz are represented in Figures 3 and 4.

Figure 3.3 depicts characteristic PD oscillograms, obtained from the measurement sensors of the 220 kV bushings of all autotransformer phases. PD distinguishing features are the steep front of signals in this case, the absence “of the prepulse” of another polarity, which characteristic for the electrostatic influence of charge of streamer corona, and, furthermore, the symmetry of PD signals on both half-waves of 50 Hz voltage, which indicates the appearance of discharge processes in the region of uniform field.

The distribution of the PD signals in the form of three-dimensional dependence “n-Q-φ” is represented in the Figure 3.5.

As can be seen from obtained PD oscillograms, in the transformer insulation of phase “B” is registered the partial discharges, which correspond to the level of amplitude in 10 nC. The form of amplitude-phase diagrams (AFD) makes it possible to assume by possible PD center in tap-changer of phase “B” [by 15-18].

According to PD Standards the maximum permissible apparent charge of partial discharges with the manufacturer tests of transformers with 150-500 kV voltage is defined as 0,3 by nC with the greatest operating voltage. For the transformers 110 kV and above for the operating conditions and aged equipment maximum magnitudes of the PD charge it is possible to accept value 3-5 nC (GOST-20074-83 “Method of measuring the PD characteristics and IEC-60270 “The measurements of PD characteristics”).

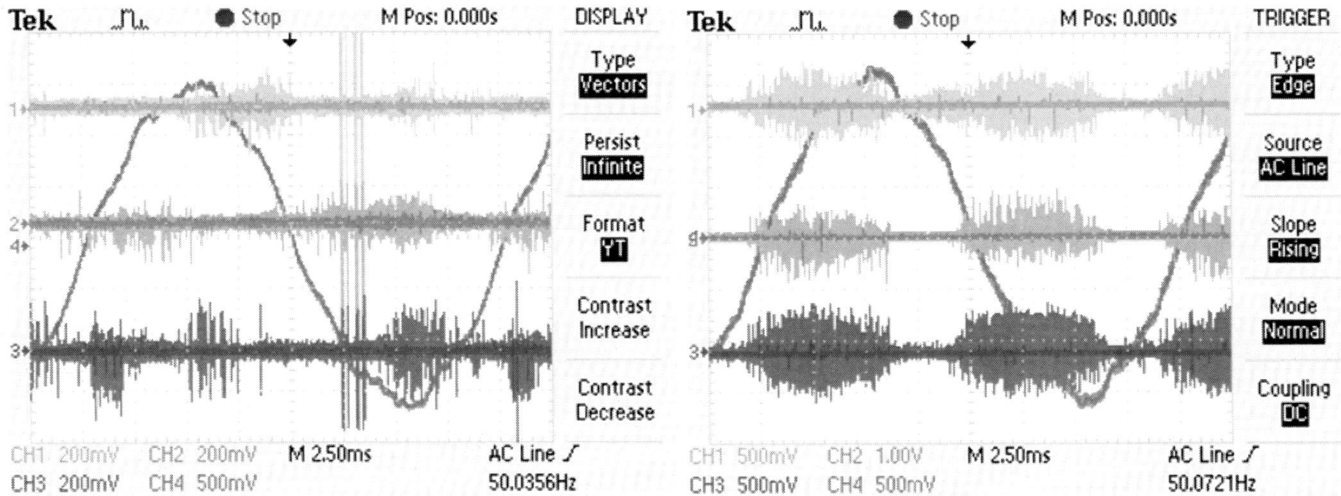

Figure 3.3. The common form of the distribution of PD signals on the autotransformer phases (side 220 kV) during several periods 50 Hz: 1 – signal from measurement bushing of 220 kV "A" phase; 2 – signal from measurement bushing of 220 kV "B" phase; 3 – signal from measurement bushing of 220 kV "C" phase; 4 - 50 Hz signal from measurement bushing of 220 kV "A" phase.

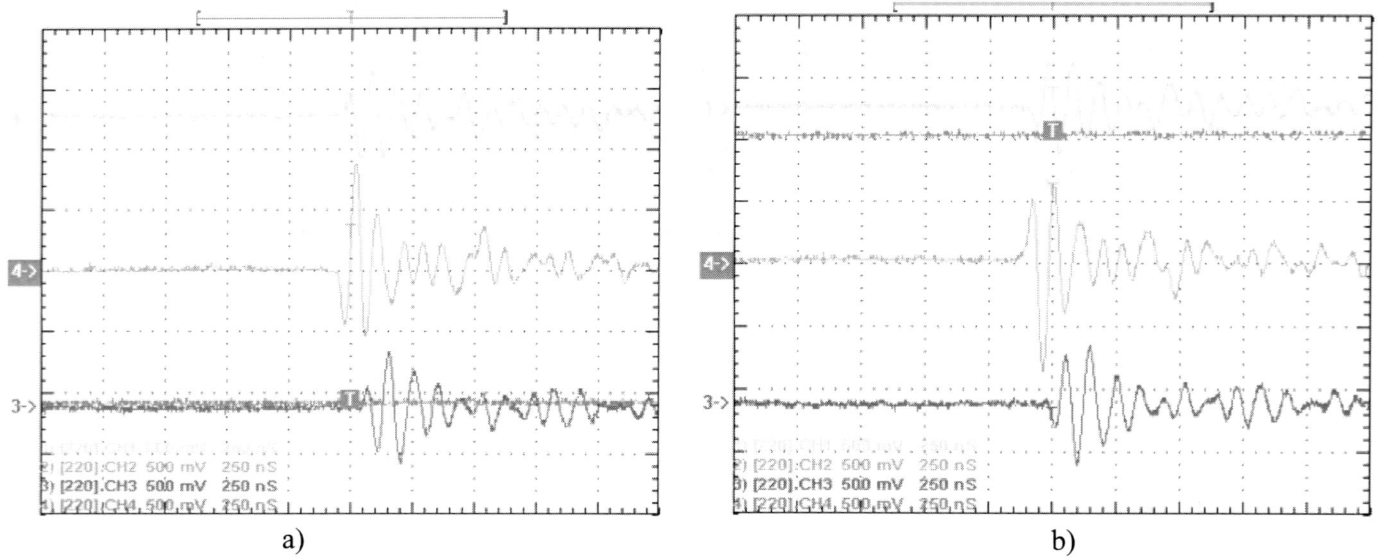

Figure 3.4. PD signals: a) 1 – signal from measurement bushing of 220 kV “A” phase; 2 – signal from measurement bushing of 220 kV “B” phase; 3 – signal from measurement bushing of 220 kV “C” phase; 4 - 50 Hz signal from measurement bushing of 220 kV “A” phase. b) 1 – signal from measurement bushing of 220 kV “A” phase; 2 – signal from measurement bushing of 220 kV “B” phase; 3 – signal from measurement bushing of 220 kV “C” phase; 4 - 50 Hz signal from measurement bushing of 220 kV “A” phase.

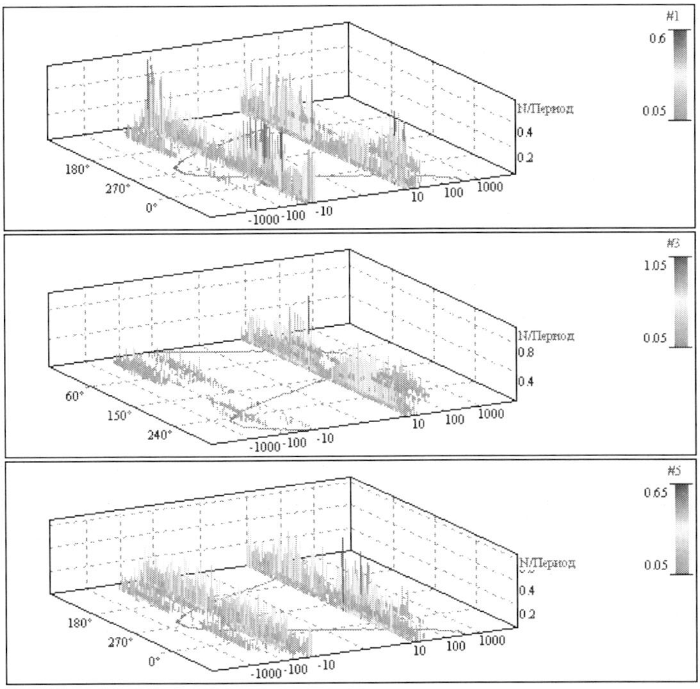

Figure 3.5. Distribution of the PD signals in the form of three-dimensional dependence "n-Q-φ", where n - a quantity of PD pulses during the period of 50 Hz frequency, Q – the PD pulse amplitude, φ - the phase angle of the appearance of PD pulses.

Therefore as a whole according to the results of inspection it was possible to estimate the state of main autotransformer insulation and bushing insulation as satisfactory. Were at the same time discovered discharge phenomena in the tap-changer device of phase "B." Therefore it was recommended to estimate the state of oil in tap-changer and, during exceeding of Standards to gas- and moisture content, the increased tangent (tan δ) for 70° and 90° temperatures, high acid number and to conduct revision of tap-changer into the spring-summer season with the replacement of transformer oil.

As a result the internal inspection of autotransformer windings after emergency switching off was revealed the damage and short-circuit of the forth turns of regulating winding (Figure 3.6).

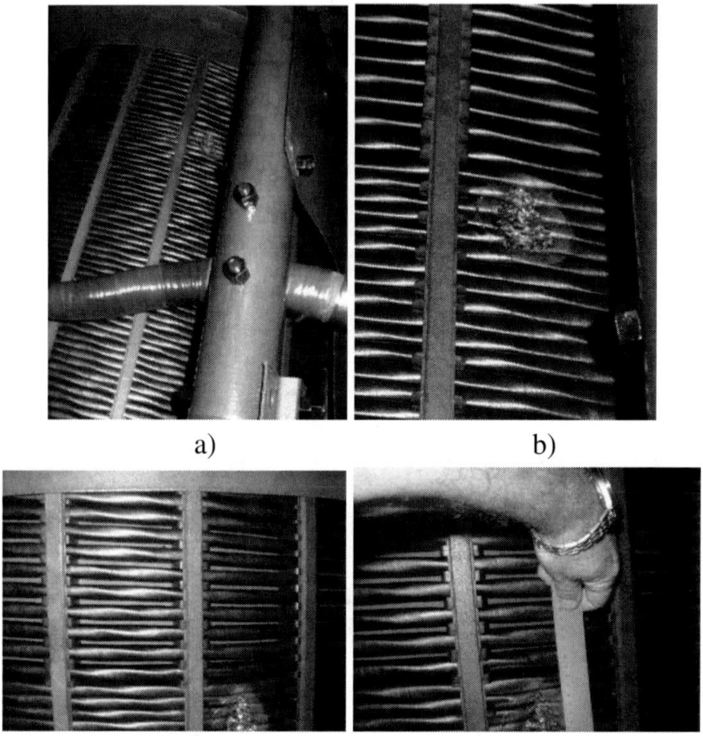

a) b)

c) d)

Figure 3.6. Internal short-circuit of 167 MVA/500/220 kV autotransformer regulating windings.

As in the report for the sums of the commission work the suppositional reason for the damage of 167 MVA/500/220 kV autotransformer was reflected the entry of moisture into the internal volume during the replacement of the bushing. This version became determining and it was proposed by the representative of transformer manufacturer [15-18].

An electrical activity of the partial discharges (PD) into the insulation of 80 MVA/110 kV power transformer are in the Figure 3.7. Color shows intensity PD per second (from the dark to the light).

Normative documentation during PD measurements are national standard GOST-20074-83 "Method of measuring the PD characteristics and IEC-60270 "The measurements of PD characteristics" [by 1-2, 3-10].

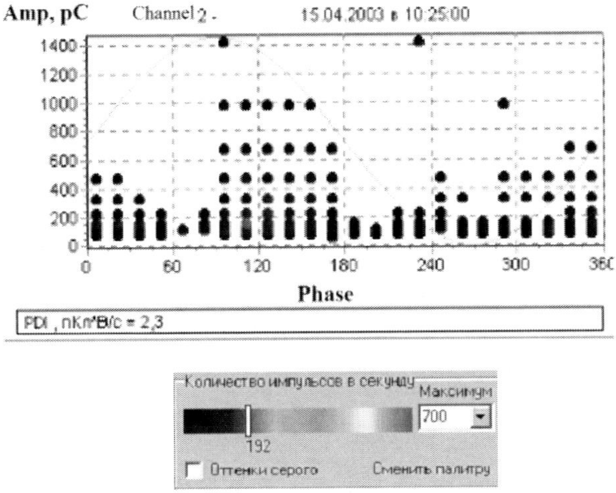

Figure 3.7. An electrical activity of the partial discharges (PD) into the insulation of 80 MVA/110 kV power transformer.

3.5. Advantages of Cables with XLPE Insulation

Cable construction with XLPE insulation is in the Figure 3.8.

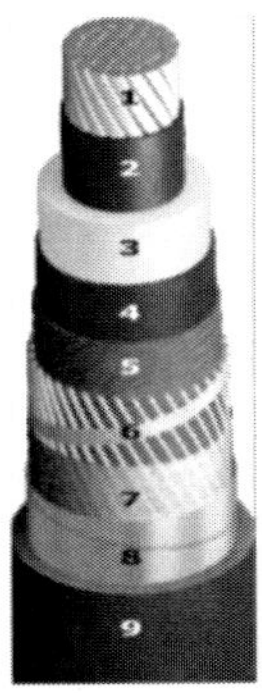

Figure 3.8. Cable construction with XLPE insulation, where is: 1. Round multiwire sealed conductor core material -Aluminum, Copper, 2. Screen for the core of extruded semiconducting cross-linked polyethylene; 3. XLPE insulation (Pv); 4. Screen for insulation from extruded semiconducting cross-linked polyethylene; 5. Separating layer 6. Screen made of copper wires fastened with copper tape of different sections: 7. Separating layer of two ribbons of crepe paper or rubberized fabric or polymer tape; 8. Separating layer 9. Sheath: polyethylene (P), polyethylene, increased thickness (Pu), plastic (B); plasticity of reduced fire hazard.

In recent years, cables with XLPE insulation have become widespread on the Russian market. The accompanying documentation provided by the manufacturers lists the advantages of cables with XLPE insulation before cables with impregnated paper insulation:

- higher reliability in operation;
- Lower costs for the reconstruction and maintenance of cable lines;
- low dielectric losses (dielectric loss factor 0.001 instead of 0.008);
- high resistance to damage;
- Greater throughput due to the increase in the permissible heating temperature of the veins: long (90C instead of 70C), with overload (130C instead of 90C);
- higher thermal stability current in case of short circuit (250C instead of 200C);
- Low permissible temperature for laying without preheating (-20C instead of 0C);
- low moisture absorption;
- Lighter weight, diameter and bending radius, which facilitates laying on difficult routes;
- possibility of laying on routes with an unlimited difference in levels;
- more ecological installation and operation (absence of lead, oil, bitumen);
- Large construction length.

Disadvantages:

- In the case of an electrical breakdown of a solid dielectric, a XLPE cable will not be able to restore its electrical strength, and any single-phase ground fault (SPGF) will lead to a stable emergency mode.
- Paper insulation cable in 60-70% (SPGF) cases is self-destructing and the operating personnel do not record these emergency modes.

Despite the advantages of cables with XLPE insulation, their share in the total production of power cables at 6-35 kV is small and has practically not grown in the last three years [1]. This is due to the fact that the main cable consumption is for repairs. And when repairing a cable line (CR), where the paper insulation is used, in most cases the cut out section is replaced with a similar one, i.e., to the cable with the paper insulation. When building new cable lines, cables with SPE insulation are mainly used. Production of power cable for voltage (6-35 kV) by type of insulation (Figure 3.9).

The structure of consumption of power cables in the directions of application is as follows:

- repair of cable lines - 55%
- reconstruction - 27%
- new construction - 18%

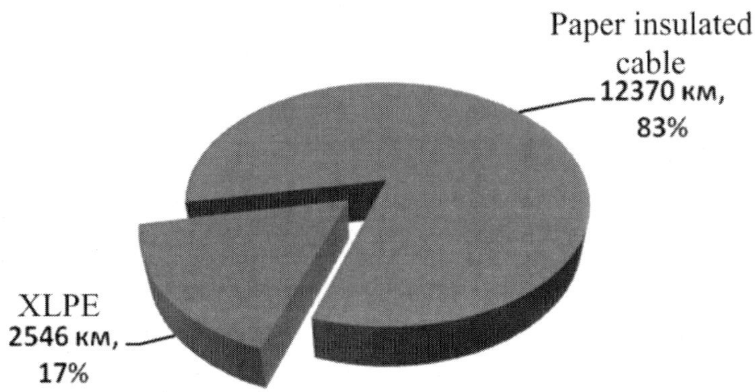

Figure 3.9. Production of power cable for voltage (6-35 kV) by type of insulation.

This explains the small share of the XLPE cable in the total production of the power cable.

During the introduction of XLPE cables in operation, design, installation and operating organizations have faced a number of problems. The main one is the lack of all-Russian normative documents for all types of work when using XLPE cables.

Already at the design stage, there are inconsistencies with the recommendations for European systems, since the cable screen section for its use in systems with grounded and isolated neutral is different (16-35 mm2 and 50-150 mm2, respectively). Also, there is no single instruction for laying a XLPE cable, each manufacturer develops it independently. It turns out that for the same brand of cable according to the technical specification there can be more than 10 instructions regulating the requirements for laying cable lines.

3.6. Comparative Characteristics of Cable with Paper Insulation and Insulation from XLPE

In general, XLPE cables are manufactured in single-core design, and the use of various types of casings and the possibility of sealing allows the cable to be used both for laying in the ground and for cable structures, including for group laying.

The XLPE cable can replace a paper-insulated cable in almost all cases, however, at the stage of introduction of cables with XLPE insulation in a particular enterprise, it is necessary to identify those areas where their application is most appropriate. To do this, we will conduct a short technical and economic comparison of traditional and XLPE cables. Unfortunately, due to differences in the costs of repairs and maintenance of cable lines for specific enterprises, the difference in overall operating costs is difficult to estimate, so we suggest comparing only the initial investment in the cable.

For a correct comparison, let's take the cables with the same throughput - paper ASB 3x240 10 kV and three single-phase APVP cables 1x185/25-10 kV. Comparative characteristics of the cables are given in Table 3.1.

It can be seen from the data given that with the same throughput and the best remaining parameters, the cost of the XLPE cable is approximately 60-70% higher. This is due to more expensive materials and manufacturing technology.

Table 3.1. Comparison between paper 3x240 10 kV and three single-phase XLPE cables 1x185 / 25-10 kV.

Comparison parameters	**Cable with paper insulation ASB 3 x 240 - 10 kV**	**Single-core cable with XLPE insulation 3 x APVP 1x185/25 -10 kV**
Cable type in section	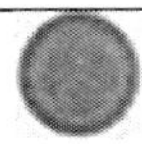	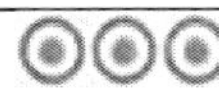
Cross-section of cores, mm2	240	185
Load current for laying in the ground, A	355	In the plane/by a triangle 375/360
Maximum permissible 1-sec short-circuit current, kA	20,56	17,5
Outside diameter, mm	62	36
Building length, m	500-600	Up to 1400
Minimum bend radius, m	1,64	0,54
Weight, kg/km	7050	1370(4110)
Permissible level difference, m	15	not limited
Comparative cost,%	100	160

This picture changes drastically as the requirements for the capacity of the cable line increase. For example, parallel cables ASB 1 x 240 10 kV should be replaced by a XLPE cable with a larger cross-section (see Table 3.2).

For the XLPE cable at 35 kV, the situation is even more favorable (see Table 3.3).

This is explained by the fact that it is impossible to apply this construction with sectoral veins to this voltage class. Therefore, paper cables are manufactured with separately leaded cores, which entails a significant rise in price compared to 10 kV cables. The cost of cables with paper and polyethylene insulation of the same cross section are approximately equal. However, as can be seen from Table 3.3, the polyethylene cable gives a 40% advantage in the load capacity.

Table 3.2. Comparison between parallel paper cables 1x240 10 kV and XLPE cable with a larger cross-section.

Comparison parameters	**Cable with paper insulation 2 x ASB 3x240**	**Single-core cable with XLPE insulation 3 x APVP 1x500/25**
Cable type in section	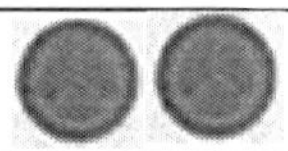	
Cross-section of cores, mm2	240	500
Load current for laying in the ground, A	639	In the plane/by a triangle 650/610
Maximum permissible 1-sec short-circuit current, kA	20,56	47
Outside diameter, mm	62	46
Building length, m	500-600	Up to 850
Minimum bend radius, m	1,64	0,74
Weight, kg / km	2x7050	2570(7110)
Permissible level difference, m	15	not limited
Comparative cost,%	100	115-120

Table 3.3. Comparison between 35 kV paper cables and 35 kV XLPE cable.

Comparison parameters	**Cable with paper insulation AOSB 3 x 150 – 35 kV**	**Single-core cable with XLPE insulation 3 x APVP 1x150/25 – 35 kV**
Cable type in section	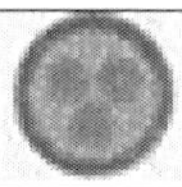	
Cross-section of cores, mm2	150	150
Load current for laying in the ground, A	250	In the plane/by a triangle 350/330
Maximum permissible 1-sec short-circuit current, kA	7,58	14,2
Building length, m	300	Up to 1000
Minimum bend radius, m	1,45	0,67

Comparison parameters	**Cable with paper insulation AOSB 3 x 150 – 35 kV**	**Single-core cable with XLPE insulation 3 x APVP 1x150/25 – 35 kV**
Weight, kg/km	6400	1805(5415)
Permissible level difference, m	15	not limited
Comparative cost,%	100	100-105

3.7. METHODS AND EQUIPMENT FOR TESTING XLPE CABLES BY VOLTAGE ON THE ULF (0.1-0.01 HZ.)

BAUR products cover all key requirements of network operators with respect to cable testing and cable diagnostics (Figure 3.10).

Underground cables have to be tested before on-site commissioning or recommissioning. Although cables are tested by the manufacturer before delivery, they maybe damaged while being laid. As well joints or terminations may be defective.

Therefore laying construction companies and network operators are obligated to carry out standard tests. In the paper insulated cables have been tested with DC voltage with the leakage current being an indication of the circuits quality.

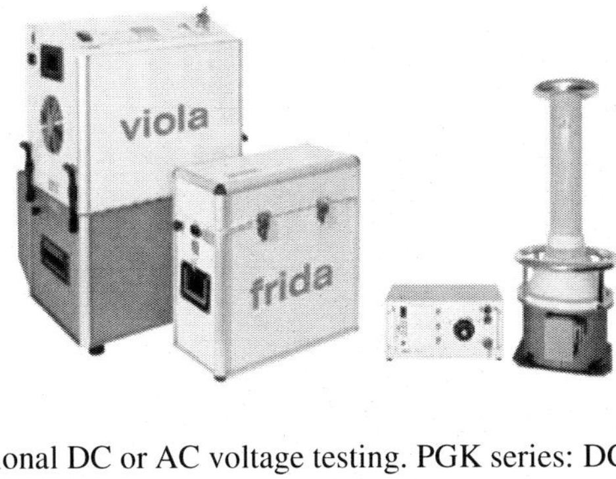

Figure 3.10. Traditional DC or AC voltage testing. PGK series: DC AC voltage up to 260 kV, AC voltage up to 190 kV (50 Hz).

Modern cables with EPR/XLPE insulation require a sinusoidal voltage (BAUR truesinus® VLF) and should be tested in accordance with the most recent national or international specifications. Such specifications define the voltage level and time for which the user performs the test. If at the end of the test time the circuit has withstood the applied voltage, then it gives the user a very high degree of confidence that when he switches it in it will operate satisfactorily.

Such a test is limited in that it gives no information about the quality of the circuit. The test result is a simple "pass/fail" information. BAUR truesinus test equipment can be used on both traditional and modern type cables.

3.8. Diagnostics of Cables with XLPE Insulation

Cable diagnostics provide the user with information about quality and condition of the circuit. This goes far beyond the "pass/fail" information from cable testing. The diagnostic information that is available with the BAUR equipment gives "present and future analysis" of the insulation status of cable systems. BAUR's cable diagnostics is used as a precautionary measure for protection against failures, it ensures optimal CBM.

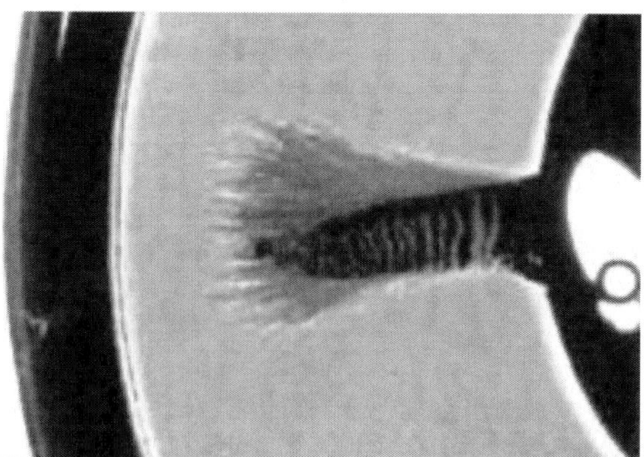

Figure 3.11. Water triing of the type "fan."

The target is to perform maintenance measures only if necessary and not to plan purely based on operating time or error probability of a cable circuit. Based on VLF test voltage, the dissipation factor measurement tan δ and partial discharge diagnostics give the necessary information without destructive testing, which can be put into a database and provide trend analysis. BAUR diagnostics technology can be used to test all cable types like plastic cables or paper-mass cables. Water triing of the type “fan” is shown at Figure 3.11.

CONCLUSION

Electron-optic detectors recommended well itself for the control of suspension isolation.

The internal short-circuits of transformer windings can be caused by the breakdown of the turn-to-turn insulation as a result of the destruction of insulation under the action of operational factors and by the action of the partial discharges in the place of future breakdown, as initiator of which can serve commutation, thunderstorm and other increased actions on the transformer insulation.

The determination of PD intensity was produced with the aid of the electrical method of PD registration. As the measuring means was used digital oscillograph, and also digital registrar-analyzer of the partial discharges.

As a result the internal inspection of autotransformer windings after emergency switching off was revealed the damage and short-circuit of the forth turns of regulating winding.

The suppositional reason for the damage of 167 MVA/500/220 kV autotransformer was reflected the entry of moisture into the internal volume during the replacement of the bushing.

Normative documentation during PD measurements are national standard GOST-20074-83 “Method of measuring the PD characteristics and IEC-60270 “The measurements of PD characteristics.”

The advantages of cables with XLPE insulation before cables with impregnated paper insulation:

- higher reliability in operation;
- Lower costs for the reconstruction and maintenance of cable lines;
- low dielectric losses (dielectric loss factor 0.001 instead of 0.008);
- high resistance to damage;
- Greater throughput due to the increase in the permissible heating temperature of the veins: long (90C instead of 70C), with overload (130C instead of 90C);
- higher thermal stability current in case of short circuit (250C instead of 200C);
- Low permissible temperature for laying without preheating (-20C instead of 0C);
- low moisture absorption;
- Lighter weight, diameter and bending radius, which facilitates laying on difficult routes;
- possibility of laying on routes with an unlimited difference in levels;
- more ecological installation and operation (absence of lead, oil, bitumen);
- Large construction length.

Cable diagnostics provide the user with information about quality and condition of the circuit. This goes far beyond the "pass/fail" information from cable testing. The diagnostic information that is available with the BAUR equipment gives "present and future analysis" of the insulation status of cable systems.

The target is to perform maintenance measures only if necessary and not to plan purely based on operating time or error probability of a cable circuit. Based on VLF test voltage, the dissipation factor measurement tan δ and partial discharge diagnostics give the necessary information without destructive testing, which can be put into a database and provide trend analysis.

ACKNOWLEDGMENTS

To Mr. Alexander G. Ovsyannikov, Roman Arbusov, Novosibirsk specialized testing laboratory, Russia, BAUR & PERGAM for valuable test equipment information, Cooperation of Universities and Innovation Development, Doctoral School project "Complex diagnostic modeling of technical parameters of power transformer-reactor electrical equipment condition," 2009 has made publishing of this article possible.

REFERENCES

[1] *IEC-60270* "The measurements of PD characteristics."

[2] *National Standard, GOST-2007-4-83* "Method of measuring the PD characteristics, 1983.

[3] Khrennikov, A. Yu. Power transformer's fault diagnostics at SAMARAENERGO Co, including FRA/LVI method. *Reports from School of Math. and System Engineering*, Vaxjo University, Sweden, N 43, 2000, ISSN 1400-1942.

[4] Khrennikov, A. Yu., Mazhurin, R. V., Radin, P. S. Infra-red and ultraviolet control, LVI-testing, partial discharges and another diagnostic methods for detection of electrical equipment's faults, defects. *Journal of Multidisciplinary Engineering Science and Technology* (JMEST) (ISSN: 3159-0040), Volume 1, Issue 4 (November 2014). http://www.jmest.org/vol-1-issue-4-november-2014.

[5] Khrennikov, A. Yu., Mazhurin, R. V., Radin, P. S. Diagnostic Methods for Detection of Electrical Equipment's Faults and Defects Including with Sulfur Hexafluoride Gas Insulation. *Energy and Power Engineering Science*, EPES 2015, 2(1): pp. 13-19. http://manuscript.sciknow.org/uploads/epes/pub/epes_1413906167l.pdf.

[6] Khrennikov, A. Yu. Smart Grid technologies for Detecting Electrical Equipment Faults, Defects and Weaknesses. *Workshop on*

Mathematical Modelling of Wave Phenomena with applications in the power industry, Linnaeus University, Växjö 23-24 April 2013. http://lnu.se/mmwp.

[7] Arbusov, R., Ovsyannikov, A. G. Application nf Optical - electronic Flaw Detector to Detect Damaged and Defective Insulators in Substation and Transmission and Distribution System. *Proceedings of CIGRE SC 33 Conference Prague September 1th - 8th,* 2000 - Paper S 4 – 2.

[8] Cimador, A., Lapeyre, J. L., Parraud, R., De Toureil, C. Reliability of insulators for overhead lines. *35th CIGRE Session.* Paris. 1994. - Panel 3 - 04.

[9] Devine, C. M., Farquhar, J. A. Bad insulators pose hidden threat. *Electrical world.* 1985. - v. 199, N12. - P. 59 -61.

[10] Thione, L. An overview of line diagnostic techniques. *38th CIGRE Session. Paris.* - 2000. - P. 1 - 02.

[11] Phillips, A. J. and Melaia, R. Ultrasonic emission from non-ceramic insulators with defects, *11th International Symposium on High Voltage Engineering,* London, UK 22-27 August. -1999. P. 131-134.

[12] Cigre Working Group 22.03 (convener C. de Tourreil). Review of'in service diagnostic testing' of composite insulators. *Electra.* 1996, No. 169, - P. 105-119.

[13] Phillips, A. J. EPRI's experiences with in-service inspection of nonceramic insulators. *Insulator 2000 World Congress*, Barcelona, Spain, November. 1999. - P. 231-240.

[14] *Symposium on Electrical Insulation*, Pittsburgh, PA USA, 5-8 June. 1994. - P. 22-24.

[15] Auckland, D. W., Smith, C. D., and Varlow, B. R. Ultrasound diagnostic tool for NDT of insulation degradation. *6th International Conference on Dielectrical Materials, Measurements and Applications.* - 1992. - P. 115-118.

[16] de Tourreil, C. Status of inspection techniques to assess the condition of non-ceramic insulators in service. *Insulator 2000 World Congress,* Barcelona, Spain, November; 1999; - P. 222-230.

[17] Vaillancourt, G. H. and Bilodeau, P. Diagnostic testing of composite insulators used on series compensation platforms in Hydro-Quebec. *11th International Symposium on High Voltage Engineering,* London, UK, 22-27 August. 1999. –P. 127-130.

[18] Spangenberg, E. and Riquel, G. In service diagnostic of composite insulators EDF's test results. *10th International Symposium on High Voltage Engineering,* Montreal, Quebec, Canada, 25-29 August. 1997, - P. 139-142.

[19] Bognar, A., Janko, A., Kreskay, E., and Szaplonczay, P. FCI Furukawa's solution for the third millenium: Enhancement in reliability of non-ceramic insulators. *World Insulator Congress.* 2001.

[20] Kuhl, M. FRP rods for brittle fracture resistant composite insulators. *IEEE Transactions on Dielectrics and Electrical Insulation.* 2001, - Vol. 8, No. 2.

[21] Bernstorf, R. A. Time-load testing of nonceramic insulators with fiberglass core rod-a 20 year summary. *1999 IEEE Transmission and Distribution Conference*, New Orleans, USA, 11-16 April. 1999. - P. 823 - 826.

[22] Burnham, J. T. and Waidelich, R. J. Gunshot damage to ceramic and non-ceramic insulators. *IEEE Transactions on Power Delivery.* 1997. - Vol. 12, No. 4, - P. 1651-1656.

[23] Scarpetta, R., Scarpetta, A., and Giobbe, G. A twenty years experience in manufacturing composite insulators. 2001 *World Insulator Congress and Exhibition, Shanghai, China*, 18-21 November. 2001. - P. 346- 360.

[24] Xidong, L., Shaowu, W., Ju F., and Zhicheng, G. Application and evaluation of composite insulators in China. 2001 *World Insulator Congress and Exhibition, Shanghai, China*, 18-21 November, 2001, - P. 29-42.

[25] Arbusov, R. S., Ovsyannikov, A. G., Lavrov, V. Yu., Tolchin, V. M. Ultraviolet and infrared flaw detection of insulating constructions. *Electric power engineer*, 2004, № 8. – P. 34-35 (in Russian).

[26] Patent № JP63175776, G 01R31.12. *Method and device for diagnosing insulation of electric equipment.* 1988.

[27] Patent № JP2003043094, G 01 R31/08. *Electromagnetic wave source detecting method, and method and apparatus for diagnosing insulation deterioration*, 1988.
[28] Patent № JP4364483, G 01 R31/08. *Detecting method of faulty or stained insulation.* 1992.
[29] Patent № JP7229948, G 01 R31/12. *Method and device for detecting defective insulation.* 1995.
[30] Patent № JP2159581, G 01 R31/12. *Judging method of deterioration of insulation of insulator.* 1990.

Chapter 4

ELECTRODYNAMIC TESTING OF REACTORS FOR RELIABILITY AND WITHSTAND TO SHORT-CIRCUIT CURRENTS

A. Yu. Khrennikov, PhD*

ABSTRACT

Chapter presents an experience of thermal and electrodynamic tests of reactors on the reliability and durability of windings during the passage of short-circuit currents serve as a tool for improving the reliability of their design and ensuring the continuity of electricity supply to electricity consumers. A prototype of 36 MVA/33 kV reactor was manufactured by transformer plant in Russia, intended for thyristor-reactor groups (TRG) of static thyristor reactive power compensators (ST RPC), was tested at the Power Short-circuit Laboratory, Togliatti during four main stages (regimes): heating tests, acoustic tests, the short circuit (SC) withstand tests (switching mode of SC and dynamic short-circuit current mode).

* Corresponding Author's E-mail: ak2390@inbox.ru.

Keywords: electrodynamic tests of reactor, reliability, thyristor-reactor groups, heating tests, acoustic tests, the short circuit withstand tests, temperature of reactor winding, distribution of currents, number of turns

4.1. INTRODUCTION

Thermal and electrodynamic tests of reactors on the reliability and durability of windings during the passage of short-circuit currents serve as a tool for improving the reliability of their design and ensuring the continuity of electricity supply to electricity consumers. A prototype of 36 MVA/33 kV reactor was manufactured by transformer plant in Russia, intended for thyristor-reactor groups (TRG) of static thyristor reactive power compensators (ST RPC), was tested at the Power Short-circuit Laboratory, Togliatti during four main stages (regimes): heating tests, acoustic tests, the short circuit (SC) withstand tests (switching mode of SC and dynamic short-circuit current mode). The known design of the cylindrical multilayer winding of the reactor with different heat transfer from the surfaces of the layers does not ensure their equal heating, which leads to the appearance of the most heated layer, uneven temperature conditions of the reactor layers and leads to a decrease in the reliability of the reactor operation. The new distribution of currents and the number of turns obtained in the article make it possible to equalize the heating temperature of all layers of the reactor winding, which reduces the temperature of the most heated point and improves the operational characteristics of the 36 MVA/33 kV reactor [1-3].

The 36 MVA / 33 kV reactor is not the only damage-prone reactor type.

One of the reasons (approximately 25%) of the winding faults of power transformers and reactors is the damage of high-voltage bushings, which in most cases leads to the spread of the accident site to the active part [4-8].

At the substation 500 kV of the West-Siberian region, 60 MVA/500 kV reactor was switched off by relay protection. There was a phase "B" damage (torn the bushing, deformation at the bottom part of the tank), gas protection was activated with the start of the fire extinguishing system. As a result of the damage, an oil was released and a fire occurred in the cell of said reactor.

A photo of the damaged 60 MVA/500 kV reactor is shown in Figure 4.1 [9-17].

Figure 4.1. Photo of the damaged 60 MVA/500 kV reactor at the 500 kV substation of the West-Siberian region.

As a result of a short circuit in the "B" phase of the reactor with subsequent ejection and ignition of the oil, the following damage occurred:

- at the 500 kV bushing, the bolted attachment of the housing of the built-in current transformers (adapter) from the tank cap of the reactor;
- the lower porcelain bushing tire is destroyed and most of the fragments fell inside the tank;
- on fragments of porcelain there are pronounced traces of decomposition of oil (yellow scurf);
- on the core of the bushing there are traces of temperature influence (blackening and charring of paper);
- break at the top of the bushing in the area of 3-4 skirts of bushing;

- displacement of the main bushing isolation with destruction of the upper bushing node;
- breaking and burning off part of the elementary conductors of the flexible wire (braids) at the output from the bottom of the bushing;
- The glass safety diaphragm is destroyed on the exhaust pipe.

The reason for the damage to the reactor bushing was the formation of the so-called "yellow scurf" on the inner surface of its porcelain bushing tire. Sedimentation occurred during operation. The presence of "yellow scurf" is difficult to diagnose by the traditional diagnostic methods.

The process of aging of the oil is accompanied by an increase in the content of the polar compounds, acids and peroxides, water and sludge. Timely replacement, or cleaning oil from aging products-a necessary condition for reliable operation of oil-filled equipment.

The main indicators characterizing the electrical insulating properties of liquid dielectrics are the permittivity ε, the tangent of the dielectric losses tan δ, the breakdown voltage U, the specific volumetric electrical resistance ρv and the gas resistance in the electric field.

Method for assessing the state of fresh and operational transformer oils on the basis of measuring their specific volume electrical resistance (ρv) and the dielectric loss tangent (tan δ) has been proposed and successfully implemented. Over the past few years more than 2000 samples of operational transformer oils have been tested, which were taken from transformers and bushings during diagnostic tests and repair work in power systems. This method is based on the measurement of specific volume resistances of oil samples in the temperature range from 20° to 90°C [1-3, 9-17].

The measurements of tan δ and ρv were performed using three-clamp cells, an electronic meter for the specific volumetric electrical resistances.

During operation, colloidal substances (resins and sludges, soaps formed as a result of the interaction of acid products of aging with metals, components of varnishes and rubbers) accumulate in the oils. Colloids are one of the possible causes of a sharp increase in the values of tan δ and a decrease in the specific volumetric electrical resistance of oils. The colloids cause the appearance of an additional loss mechanism associated with relaxation polarization in an alternating electric field.

In a number of cases, with satisfactory values of the acid number, a small amount of water and particles of mechanical impurity, a normal flash point and other parameters that correspond to the norms, the oil has high values of the dielectric loss tangent that do not correspond to the measured values of the specific volume electrical resistance [16-25].

Another example is the occurrence of a damage of 60 MVA / 500 kV reactor of another type. Cause inter-turn breakdown of reactor winding insulation due to manufactory defect (Figure 4.2).

Figure 4.2. Inter-turn breakdown of 60 MVA / 500 kV reactor winding insulation due to manufactory defect.

4.2. 36 MVA/ 33 kV Reactor Testing

The line of the reactors of the different destination and modification were tested besides the power transformers at Short-circuit Laboratory SCL, Togliatti. The experimental model of 36 MVA/33 kV reactor was intended for the thyristor-reactor group (TRG) of static thyristor compensator of reactive power (STC RP). It passed the testing at SCL in course of four main stages (modes): the temperature rise test, the acoustic test, short circuit (SC) withstand test (SC commutation mode and dynamic SC current mode). The reactor was represented the two air core winding (half). Both reactor halfs connected in the operation consecutively through the thyristor switch.

5 layers of the reactor half were connected in parallel. In all five reactor layers were used the wire of sort PAETR of same cross-section, with subsequent rolling to receive the same axial dimension of layers by different turn numbers. The main parameters of the tested reactor are in Table 4.1.

Table 4.1. Main parameters of 36 MVA/33 kV tested reactor.

Value name	Measurement unit	Value
1. Rated power	kVA	36000
2 . Rated voltage	kV	31,1
3. Rated current	A	1100
4. Rated inductance	mH	90
5. Rated inductance of the half	mH	37,4
6. Rated inductive resistance	Ohm	28,274
7. Designed exceed of temperature of the half	°C	67,7
8. Rated frequency	Hz	50
9. Main loss by 115°C	kW	132
10. Heat resistance class	-	F 155°C
11. Type of cooling	natural air	
12. Quantity of the half	-	2
13. Quantity layers in the half	-	5

4.3. Heat and Acoustic Reactor Testing

The temperature rise test of reactor was carried out by the method of direct load in accordance with demand of National Standard GOST 3484.2-88. The heating was carried out through rated current of 1100 A with a voltage 16,5 kV and a frequency 50 Hz for at least two hours.

Measurement of the winding temperature was carried out through the thermocouples using after heating reactor to the steady state thermal state [1-3, 16-25].

During tests was determined the medium temperature of the winding half, the temperature of the winding hot-spot, the temperature of reactor contact joints, the temperature of the flanges of support insulators, the temperature of frame elements, the temperature on channel output of the top reactor half, the environ air temperature. The tests were carried out by the reactor location on open air in natural weather conditions. The testing scheme of 36 MVA/33 kV reactor set on Figure 4.3.

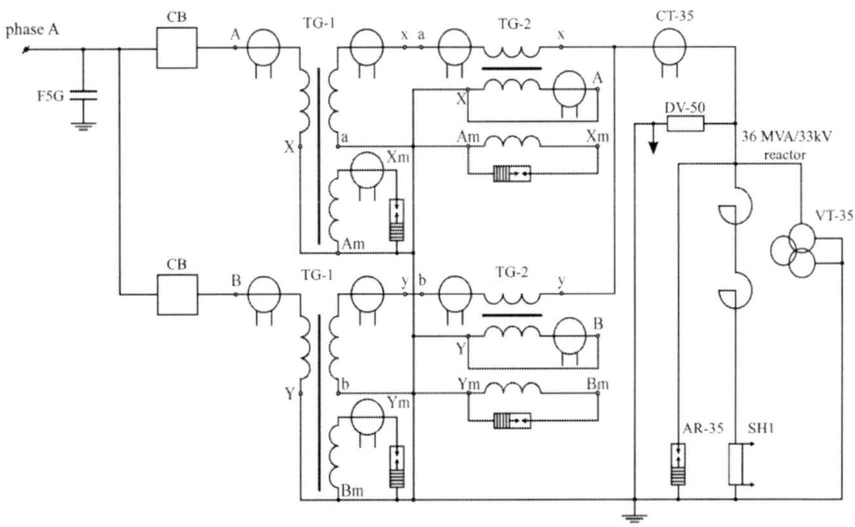

Figure 4.3. Testing schemes of the temperature rise test and acoustic test of 36 MVA/33 kV reactor, where: CB – circuit breakers; F5G-harmonic filter; TG-1, TG-2-intermediate transformer groups; DV-50 – capacitive voltage dividers; 36 MVA/33kV reactor-tested reactor; SH1 – the low-induction current-measuring shunt; CT-35 – current transformer; VT-35 – voltage transformer; AR-35-arrester.

In carrying out the temperature rise test of reactor in the course of first adjustment test duration on 0,5 s for decrease of current in the reactor was included only the one parallel circuit of the intermediate transformers – by phase "a" groups TG-1 and TG-2. The circuit of the intermediate transformers by phase "в" (TG-1 and TG-2) was in no load by the excitation from the reactor side [1-3, 27-29].

Designed current value in reactor by temperature rise test – 974 A, real current value in reactor was in first period 960 A. In the first layer current was 160,5 A, in the second layer was 188 A and so on. After first period of 1st adjustment test is occurred the electrical ceiling between the second and third layers of the top reactor half. Current value was 1604 A, that correlate of 146% from rated value.

Triggering of arresters not fixed. At reactor inspection was revealed the track of the arc burning on the top half in location of third field from jumper clockwise, crater of arc burning on metallic plate of the lower support construction of the top half between second and third layers.

On second layer of top reactor half on outer surface in third field clockwise from jumper between two reactor halfs were revealed the tracks of intensive arc burning, brackage (overburning) bandage tape in the zone between two textolit laths. On a lower support insulator-extensive soot layer (Figure 4.4) [80-83].

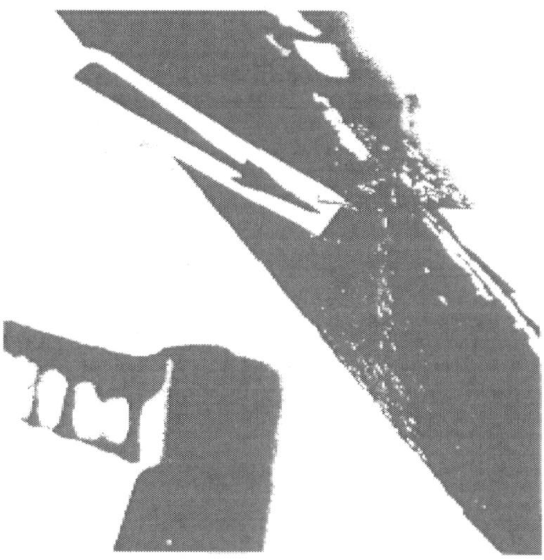

Figure 4.4. The photo, illustrating the tracks of the arc burning on second layer of top half of 36 MVA/33 kV reactor, arising in a result of the insulation ceiling on account of factory defect.

Possible reason of electrical damages can be a factory defect and a hit of the metallic details in space between second layer and third layer of the top half of 36 MVA/33 kV reactor.

Further the temperature rise test were continued on lower reactor half. In all were carried out 1 adjustment test temperature rise tests of duration 1 hour and 3 credible test of a duration 2 hours each.

The excitation set on top part of lower half reactor by the simultaneous earthing of the lower part, where is located the temperature sensors: top flange of lower insulator, outer and inner layer of lower part of lower half, lower cross.

The temperature of the environ air in the course of the reactor temperature rise tests was 0 ÷ 3°C [1-5, 27].

The carried out control measurements on reactor have indicated the inductance changing of top half at +0,14%, lower half at +0,08%, the ohmic resistances of second and third layers of top reactor half increase at 80 ÷ 87%.

In the course of adjustment test temperature rise test of lower half reactor voltage was 13,9 kV, current value was equal to 1110 A.

After opening of the reactor was the temperature of insulator flange 107°C, outer 5-th layer of reactor 59°C, inner 1-st layer 20 ÷ 25°C, the cross 11°C.

In the course of subsequent three credible tests the current value was 1098 ÷ 1104 A. That is equal 100,3% from normative value, voltage-13,52 kV.

Schedules of changing temperature of divers points of reactor by carrying out of temperature rise test 1st credible test и changing of the current, flowing through the reactor winding, set on Figure 4.5.

According to National Standard GOST-8865-70 reactor is considered stood the temperature rise test, if a relative temperature exceeding его winding over temperature of the environ cooling medium, determined by formula not exceed the value 100°C for the heating resistance class F:

$$\Delta\theta win = \theta win - \theta air, \tag{4.1}$$

where θwin - the winding temperature in registration point temperature; θair - the temperature of outer cooling medium (environ air).

Maximum exceeding of the temperature of winding of 36 MVA/33 kV reactor over temperature of environ air in the course of temperature rise test was 91°C for the outer 5th layer of winding. This is associated with highest current density 1,7 A/mm2 of outer layer, and also due to constructively the outer layer is located next to the insulator flanges, which exert the mutual heating influence. The current density of the inner layers was 1,1 ÷ 1,4 A/mm2.

The received exceeding 91°C temperature winding of 36 MVA/33 kV reactor is in the limits, allowable for reactor class of heating resistance F (155°C). Therefore 36 MVA/33 kV reactor for ST RPC can consider stood the temperature rise test.

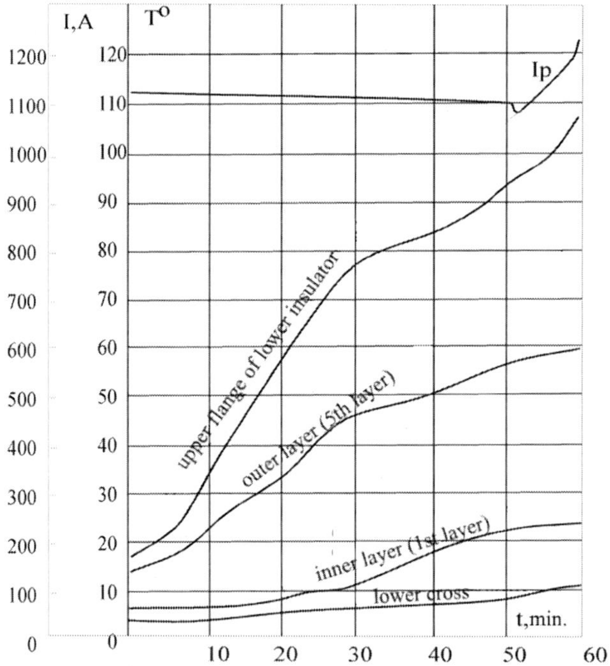

Figure 4.5. The schedules of the changing of the temperature and current of the lower half of 36 MVA/33 kV reactor in the course of the temperature rise test.

The medium level sound of dry 36 MVA/33 kV reactor in carrying out the acoustic testing was on distance 4 m 65,6 dBA, that corresponds to the norm.

Exceeding temperature of contact joint, the current conductive parts (frame) and nonconductive parts (posts) over environ temperature exceeds allowable value, equal for contact connection 50°C, end for others 80°C according to National Standard GOST 8024-90 [1-3, 27, 30-42].

By the study of results of temperature rise test reactor и well-known design of cylindrical multilayer winding from wire with rectangular cross-section for air core reactor was made a conclusion, that well-known designs by divers heat irradiation с layers surfaces winding reactor do not provide with the same temperature rise of all layer of the winding [1-3, 43].

It leads to appearance of most heated layer, uneven temperature mode of a reactor layers. It leads in the end to reliability decrease of the reactor work.

4.4. The Calculation of the Current Distribution

In the well-known devices [43, 44] cylindrical winding consists of in parallel connected concentric layers. The turn number in layers increase from inner to outer, and all layers of winding wrapped around same wire of same cross-section.

The turns number of layers from wind to layer, with all layers of winding wrapped around the same wire of the same cross-section.

The layers, following за inner, consist of two parts. In first из которых wire in turn is located small side rectangular cross-section parallel to the vertical winding axe, a во second-larger side of the rectangular cross-section.

The turn quantity in the second part в each layer increases to outer layers and equal:

$$W_{n_n} = \frac{H - W_n\ aT}{(b-a)T}, \tag{4.2}$$

where W_{n_n} - the turn number of the nth layer, in which a wire located большей side сечения in parallel to the vertical axe winding;

W_n - total turn number in nth layer;

H – winding height;

a-length of the smaller side of the cross-section of the rectangular wire;

b-length of the larger side of the cross-section of the rectangular wire;

T – factor of the wire winding tightness;

n – the layer number [1-3, 27, 43-44].

Farther for the simplify of the calculation of the current distribution set the assumption of same heat of the all layers of reactor winding by the different heat irradiation с их surfaces. It allows for the losses distribution by layers to use a following condition:

$$\Delta\theta_n = idem, \tag{4.3}$$

where θ_n - exceeding temperature wire of the nth layer over temperature of the cooling air.

From the thermodynamics is know the equation associated layers the loss power с exceeding of temperature between the cooled surface and cooling medium:

$$P_n = I_n^2 \cdot \rho \cdot \frac{\pi \cdot D_n \cdot W_n}{S_{bn}} \cdot k_{gn} = \alpha_{n1} \cdot \pi \cdot D_{n1} \cdot h \cdot \Delta\theta_n + \alpha_{n2} \cdot \pi \cdot D_{n2} \cdot h \cdot \Delta\theta_n \approx$$

$$\approx (\alpha_{n1} + \alpha_{n2}) \pi D_n h \Delta\theta_n, \tag{4.4}$$

where I_n -the current in the nth layer;

D_n, D_{n1}, D_{n2} - diameters of the nth layer, medium, inner and outer accordingly, practically same $D_n \approx D_{n1} \approx D_{n2}$;

α_{n1}, α_{n2} -the heat irradiation factors c inner and outer surfaces of the nth layer;

S_{bn} - the active cross-section of the wire of the nth layer;

h – the layer height;

k_{gn} -the additional loss factor.

On the base (5.23) we will find the ratio between the currents in layers:

$$\frac{I_1}{I_n} = \frac{\sqrt{(\alpha_{11}+\alpha_{12})/W_1}}{\sqrt{(\alpha_{n1}+\alpha_{n2})/W_n}} = \frac{\sqrt{\alpha_1/W_1}}{\sqrt{\alpha_i/W_n}}, \quad (4.5)$$

где $\alpha_i = \alpha_{n1} + \alpha_{n2}$ -the total heat irradiation factor of the nth layer;

W_n -the relative turn number in nth layer.

При этом the heat irradiation factors of layers will be equal:

- for the inner surface:

$$\alpha_{n1} = \frac{3,5}{\sqrt[4]{h}}(\Delta\theta_n)^{1/4}, \quad (4.6)$$

- for the outer surface:

$$\alpha_{n2} = \frac{5}{\sqrt[4]{h}}(\Delta\theta_i)^{1/4} \quad (4.7)$$

The above received relations answer to the condition of the same heating of the all layers of the cylindrical winding for the reliability rise of the reactor in the operation.

On the base of formulae (4.1)-(4.7) can to build the new design of cylindrical winding for the dry reactors c on ensuring of the named above same heat of all layers of the reactor winding. The winding consists of concentric layers, connected in parallel, from the inner layer and subsequent layers.

The turn numbers in a layers is divers and decrease from inner layers to outer for the sake of ensure the equality currents and heat temperature in parallel layers in accordance with expression (4.5).

The equality (4.5) in essence is the condition of the same heating of the all layers of cylindrical winding. It eliminate the winding local overheating, increase works term and operation reliability of a dry electrical reactor.

The calculations of the turn number and distribution of currents by layers we will consider on the example of the winding of 36 MVA/33 kV reactor.

Table 4.2. Parameters of cylindrical winding of the compensation of 36 MVA/33 kV reactor

Height of half, mm		**1340**
Layer number		**5**
Required total turn number in layer (Wn):		
1	(W1)	104,3
2	(W2)	99,4
3	(W3)	97,5
4	(W4)	93,6
5	(W5)	93,6

Table 4.3. Calculation turn number and current distribution by layers

Number of layer		1	2	3	4	5
Factor of heat transfer with surface	α_{n1}	3,25	1,63	1,63	1,63	1,63
	α_{n2}	1,63	1,63	1,63	1,63	4,64
	α_n	4,88	3,26	3,26	3,26	8,27
Relative turn number	W_n	1,10	1,05	1,02	1,00	1,00
Current	I_n	230	195	200	200	275

The cylindrical winding for a single phase compensation reactor in dry with air core for the STC RP of the voltage class 35 kV power 36 MVA of 36 MVA/33 kV reactor has following design parameters (Table 4.2).

According to formylae (4.1) and (4.5) we will calculate the heat irradiation factors of the surface α_{ni}, relative turn number and current of the nth layer according to the formulae (4.5). The calculation results set in 3.

The received current distribution and turn number allow to make even the heat temperature of all layers of reactor winding. It reduces the hot-spot temperature and improves operation characteristics of 36 MVA/33 kV reactor [1-3, 27-29, 32-38, 43-44].

4.5. Reactor Testing for Withstand to the Aperiodical Overshooting of Load (Commutation Short-Circuit) and Electrodynamic Short-Circuit Current

The load current impact thrust withstand tests (commutation SC) and the dynamic SC currents withstand tests.

The tests in mode of commutation SC currents and the dynamic SC currents withstand tests were carried out on the lower reactor half by simultaneously short-circuited top half.

36 MVA/33 kV reactor stood the load current impact thrust withstand tests (commutation SC). It were carried out 10 thousand impact of value currents 2I*nom* (nominal value).

Overvoltage level in the tests was (1,12-1,15)U*nom*. The total reactor currents was in a limit of 2,173 kA ÷ 2,262 kA. It is 96,4% ÷ 102% from normalized current value.

36 MVA/33 kV reactor has stood also the SC withstand tests in mode of the dynamic SC current. In mode of the electrodynamic SC were carried out 5 credible tests.

The impact SC currents values was from 7,726 kA to 8,417 kA, that is, from 94,7% to 104% from the normalized value. The steady state SC current values was from 2,960 kA to 3,174 kA. It correlate 92,5% ÷ 99,2% from the normalized value.

In the course of carry out of all test modes tests of 36 MVA/33 kV reactor were carried out the control of the reactor winding condition. It is used the measurement method of the active loss by the power supply on both reactor half and on each separately, a also the method of measurement SC reactor reactance at lowering voltage.

Maximum deviation value of the reactance by the power supply on both halfs was +0,14% by measurement after the temperature rise test, in the course during which the electrical breakdown occurred on the top reactor half (Figure 4.4).

Tests in the modes of the commutation current SC and dynamic current SC is carried out on lower reactor half by the simultaneously shorted top half.

36 MVA/33 kV reactor has stood the load current impact thrust withstand test (commutation SC). It were carried out 10 thousand impacts by the current nominal value 2I*nom*.

The overvoltage level in the tests was (1,12 ÷ 1,15)Ur. The total winding current was in a limits 2,173 kA ÷ 2,262 kA. It consists 96,4 ÷ 102% from normalized current value [1-3, 27-40, 45-49].

CONCLUSION

1. 36 MVA/33 kV reactor passed the testing at SCL in course of four main stages (modes): the temperature rise test, the acoustic test, short circuit (SC) withstand test (SC commutation mode and dynamic SC current mode).
2. Maximum exceeding of the temperature of winding of 36 MVA/33 kV reactor over temperature of environ air in the course of temperature rise test was 91°C for the outer 5th layer of winding. This is associated with highest current density 1,7 A/mm2 of outer layer, and also due to constructively the outer layer is located next

to the insulator flanges, which exert the mutual heating influence. The current density of the inner layers was 1,1 ÷ 1,4 A/mm2.

3. The known design of the cylindrical multilayer winding of the reactor with different heat transfer from the surfaces of the layers does not ensure their equal heating, which leads to the appearance of the most heated layer, uneven temperature conditions of the reactor layers and leads to a decrease in the reliability of the reactor operation.
4. The new distribution of currents and the number of turns obtained in the article makes it possible to equalize the heating temperature of all layers of the reactor winding, which reduces the temperature of the most heated point and improves the performance of 36 MVA/33 kV reactor.

ACKNOWLEDGMENTS

To Dr. Oleg Shlegel, scientific supervisor of my master's (candidate) thesis; Mr. Nikolaj M. Alexandrov, post-graduate student, Russia, JSC "Dinamika," Cooperation of Universities and Innovation Development, Doctoral School project "Complex diagnostic modeling of technical parameters of power transformer-reactor electrical equipment condition," 2009 has made publishing of this article possible.

REFERENCES

[1] IEC 60076-2, *Power transformers – Part 2: Temperature rise for liquid-immersed transformers*. 2011, 100 p.

[2] IEC 60076-3, CEI 60076-3. *Power transformers – Part 3: Insulation levels, dielectric tests and external clearances in air*. 2013, 124 p.

[3] IEC 60076-5. *Power Transformers – Part 5: Ability to withstand short circuit*. 2006, 80 p.

[4] Verhoeven, B. Shunt reactor current tests with HV and UHV circuit breakers. More routine based switching of lower current can present significant concerns in distribution and transmission networks. *DNV GL's Quarterly Newsletter of Testing, Inspections & Certification.* 2016. https://www.dnvgl.com/article/shunt-reactor-current-tests-with-hv-and-uhv-circuit-breakers-76628.

[5] Wang, M. Y. (2001). 1995-1999 Fault Statistics & Analysis. All Transformer Types in China. *Electrical Equipment*, 49(1), p. 11-12.

[6] IEC 60909-3:2009. *Short-circuit currents in three-phase AC systems-Part 3: Currents during two separate simultaneous line-to-earth short circuits and partial short-circuit currents flowing through earth.*

[7] Khrennikov, A. Yu., (1999). *Control and protection device of transformer windings against deformation at short circuits*, Patent of Russian Federation, № 2136099. Bulletin № 24.

[8] S. Bharadwaj Reddy. *Current Limiting Reactors and Short Circuit Currents.* https://instrumentationtools.com/current-limiting-reactors-and-short-circuit-currents/.

[9] Khrennikov, A. Yu., Shlegel, O. A., Lurie, S. I., (1996). *Control and protection device of transformer windings against deformation at short circuits in service.* Patent of Russian Federation, № 2063050. Bulletin № 18.

[10] Khrennikov, A. Yu., (1998). *Measuring device of the inductive transformer short-circuit impedance*, Patent of Russian Federation, № 96110718 A.

[11] Salama, M. M. A., Dudley, R. F., Gela, G. Calculation of steady-state surface stresses at winding ends of air-core power reactors. *IEEE Proceeding*, 1981.

[12] Fogelberg, T., Girgis, R. S., (1994). ABB power transformers - a result of merging different technologies with prospects for significant future advancements," International Simposium *Electrotechnics-2010*, Moscow, vol.1, 56-59.

[13] Balakrishnan, V., (1989). A study of short-circuits in large power transformers, *Electrical India*, Vol. XXIX, № 4, 87-91.

[14] McNutt, W. J., Johnson, W. M., Nelson, R. A., Ayers, R. E., (1970). Power transformer short-circuit strength-requirements, design, and demonstration. *IEEE Trans. Power Appar. and Syst.,* 89, № 8.

[15] Malewski, R., Khrennikov, A. Yu., Shlegel, O. A., Dolgopolov, A. G. (1995). Monitoring of Winding Displacements in *HV Transformers in Service, Reports of CIGRE Working Group 33.03.* Italy, Padua.

[16] Khrennikov, A. Yu., (1999). Short-circuit performance of power transformers. Test experience at Samaraenergo Co and at Power Testing Station in Togliatti, including fault diagnostics, *Reports of CIGRE Study Committee 12 Transformers*. Hungary, Budapest.

[17] Khrennikov, A. Yu., (2000). Power transformer's fault diagnostics at Samaraenergo Co, including FRA/LVI method, *Reports from School of Math. And System Engineering*, Vaxjo University, Sweden, № 43, ISSN 1400-1942.

[18] Wang, M. Y. (2006). 2001-2005 Fault Statistics & Analysis of All Transformer Typesin China. *Electrical Devices*, 37(2), p. 4-13.

[19] Zhang, K. L. (2001). Improve the ability of large power transformer short-circuit resistance measures. *Fujian Dianli Yu Diangong*, 1(3), p. 8.

[20] Zhang, F. J. & Wu, Z. Y. (2009). Application actualities and development of short-circuit limiting technique for power system. *Guangdong Electric Power*, 22(2), p. 11-14.

[21] Han, L., Qiu, W. D., Xiao, Zh. L.(2009). Overview on measures of short-circuit limiting technique on power system. *Electric Power Technologic Economics,* 21(3), p. 33-37.

[22] Lin, Y. (2010)Application of short-circuit current limitation technologies for 500kV power system in Guangdong Grid. *North China Electric Power, Mechanical & Electrical Engineering Techbology,* 39(9), p. 97-100.

[23] ABB_Australia. (2006). *ABB Capability Guide* http://www04.abb.com/global/seitp/seitp202.nsf/c71c66c1f02e6575c125711f004660e6/57af7451ece31b4c482573ea001f8711/$FILE/AUABB%20Capability%20Guide_LR.pdf.

[24] Zhang, L., Guo, Q., Zeng, L. D. & Dai, C. (2011). High-Capacity Explosive High-Speed Interrupted Mechanism for Independent Power System. *Journal of Naval Aeronautical and Astronautical University*. 26 (5), p. 561-566.

[25] Han, G., Han, L., Wu, L. (2010)Application and development of methods on limiting power grid's short-circuit current. *Power System Protectionand Control* 38(1), p. 131-144.

[26] He, Zh. Q., Gao, W. L. (2010). Analysis on restrictive measures against short-circuit current of 220kV Chongqing power grid. *Central China Electric Power*, 23(2),p. 31-34.31.

[27] Khrennikov, A. Yu., Shakaryan, Y. G., Dementyev, Y. A., (2013). Shortcurrent Testing Laboratories. Short-Circuit Performance of Power Transformers, Transformer Testing Experience. *International Journal of Automation and Control Engineering* (IJACE), Volume 2, Issue 3, 120-127. http://www.seipub.org/ijace/AllIssues.aspx.

[28] Bertagnolli, G., (1998). Short-Circuit Duty of Power Transformers. *ABB Transformatori*, Legnano (Milano) – Italy, 1997.

[29] Janssen, A. L. J., te Paske, L. H., Smeets, R. P. P., Leufkens, P. P., Fogelberg, T., (2009). YEARS TEST Experience with short-circuit withstand capability of large power transformers. *CIGRÉ 2009, 6th Southern Africa Regional Conference*, Paper P501.

[30] Abdallah, J., (2009). Using the Frequency Response Analysis (FRA) In *Transformers Internal Fault Detection. WSEAS Transactions on Power Systems*, Issue 9, Volume 4, 297-306.

[31] Khrennikov, A. Yu., (2012). New "intellectual networks" (Smart Grid) for detecting electrical equipment faults, defects and weaknesses. *Smart Grid and Renewable Energy*, Vol. 3, No 3, 159-164. http:// www.scirp.org/journal/sgre.

[32] Khrennikov, A. Yu., (20130. Fault Detection of Electrical Equipment. Diagnostic Methods. *International Journal of Automation and Control Engineering* (IJACE), Vol. 2, No. 1, 19-27. http://www.seipub.org/ijace/Archive.aspx.

[33] Khrennikov, A. Yu., (2013). Smart Grid technologies for Detecting Electrical Equipment Faults, Defects and Weaknesses. *Workshop on*

Mathematical Modelling of Wave Phenomena with applications in the power industry, Linnaeus University, Växjö, 23-24. http://www.lnu.se/mmwp.

[34] Lapworth, J. A., (2001). Recent Developments Relating to the Detection of Winding Movement in Transformers by Frequency Response Analysis (FRA). *Sixty-Seventh Annual International Conference of Doble Clients.*

[35] Lech, W., and Tyminski, L., (1966). Detecting transformer winding damage-the low voltage impulse method. *Electrical Review*, № 18.

[36] IEEE Std C57.12.90-1993, *Part II: IEEE Guide for Short-Circuit Testing of Distribution and Power Transformers.*

[37] Janssen, A. J. L., te Paske, L. H., (2000). Short-Circuit testing experience with large power transformers, *CIGRE SC12 Session, report* 12–105.

[38] Janssen, A. L. J., te Paske, L. H., Smeets, R. P. P., (2002). Short-circuit Tests on Large Transformers: Destructive or not? *TrafoTech*, Mumbai, India, 126-33.

[39] Janssen, A. L. J., te Paske, L. H., van der Linden, W. A., Smeets, R. P. P., (2001). Magnetic Saturation at Short-Circuit Tests on Power Transformers," *IEEE PES Summer Meeting*, paper 72.03.

[40] IEC Committee Draft 14/465/CD Annex B, (2003). *Theoretical Evaluation of the Ability to Withstand the Dynamic Effects of Short Circuit.*

[41] CIGRE WG12.05, (1983). An International Survey on Failures in Large Power Transformers in Service, *Electra* 88, 21–37.

[42] Bertagnolli, G., (1999). Results of short-circuit performance of transformers, CIGRE Transformer Colloquium, Budapest, Short-circuit performance; *Tests and Failures, Report 2, CIGRE TC 12.*

[43] *Patent of Austria № 374037.* H 01 F 27/28. 1984.

[44] *Patent of USSR № 1376126.* Bul. № 7, 1988.

[45] Khrennikov, A. Yu., (2009). *Complex diagnostic modeling of technical parameters of power transformer-reactor electrical equipment condition.* PhD Thesis, Samara State University (in Russian), 405.

[46] Malewski, R., Anderson, W. I., Moore, W. J. Interlaboratory comparision of EHV shunt reactor loss measurements. *Transaction on Power Apparatus and System*, Vol. PAS-99, No. 4 July/Aug 1980.
[47] DEP 33.65.40.31-Gen. Power transformers (Amendments/ supplements to IEC 60076-1 and IEC 60076-11). *Design and Engineering Practice*, 2013, 31 p.
[48] European Standard. EN 50588-1. Medium power transformers 50 Hz, with highest voltage for equipment not exceeding 36 kV-Part 1: *General Requirements*. 2015, 29 p.
[49] International Standard IEC 60076-10. *Power Transformers – Part 10: Determination of Sound Levels*. 2001, 7 p.

Chapter 5

TECHNOLOGICAL BREAKDOWNS WITH THE DAMAGE OF POWER TRANSFORMERS BECAUSE OF THE ACTION OF SHORT CURRENT CURRENTS

A. Yu. Khrennikov[*], *PhD*

ABSTRACT

The main types of defects and damages and the factors leading to their occurrence are discussed in this chapter. Constructive errors in the manufacture of transformers and reactors at the plant, defects in one of the nodes can also cause damage. A classification of probable defects causing damage of transformer-reactor equipment is made in terms of reducing resources and safety reserves.

---

[*] Corresponding Author's E-mail: ak2390@inbox.ru.

Keywords: classification of defects, reliability, transformers, heating tests, insulation, chromatographic analyze, water-oil emulsion, radial buckling, circuit breaker

5.1. Introduction

The probable defects of possible damage of transformer-reactor equipment in the insulation system, the electromagnetic system of the transformer, the current-carrying and mechanical system of transformers and reactors, as well as the classification of the main types of impacts on the transformer-reactor equipment.

It is noted that transformers with defects in the active part can be operated for many years, although in the place of the defect there will be heating, an increase in the level of partial discharges (PD) in isolation and deterioration in the results of diagnostic measurements (chromatographic analyzes, etc.). In the future years of operation, as well as in the case of the next short-circuit with the presence of an aperiodic (shock) component of the through-current fault, an emergency failure of the transformer with severe consequences is likely [1-4].

5.2. The Reason of the Damage of the Power Transformers

Carrying out the classification, we will consider the first cause of damage to power transformers and reactors damage to high-voltage bushings, which in most cases leads to the spread of the accident site to the active part of the electric. Internal of windings due to breakdown of insulation due to aging and the effect of the PD will be considered the second reason. The third reason is insufficient electrodynamic stability of windings at short circuit. External faults at the transformer terminals can also lead to the development of accidents with severe consequences appliance [3-11].

5.3. Appearance of the Winding Electrodynamic Deformations of the Power Transformers

250 MVA 500 kV/110 kV autotransformer (manufactured in 1979) was switched off by the action of differential and gas protection at the Ural region. During the audit of autotransformer with a lifting of the tank, radial deformations of 500 kV winding were detected (Figure 5.1 a, b). At the same time, the gas in the gas relay turned out to be combustible, autotransformer tank was deformed in the upper part. All diagnostic characteristics of insulation, autotransformer oil analyzes did not exceed the Standards [1-4].

a)

b)

Figure 5.1. Radial buckling in the HV winding (a) and lodging of the winding wire (b) winding 500 kV of 250 MVA 500 kV/110 kV autotransformer at the Ural region.

The most likely reasons for the failure of this 250 MVA 500 kV/110 kV autotransformer were the presence of a local defect in the high voltage winding, which was not detected by routine diagnostic methods. It is also possible to form the "lenses" of the water-oil emulsion in the lower part of the tank, and then get moisture into the windings when the AT is switched on and the oil circulation system after a long time in the cold reserve. Such defects, associated with the humidification of the winding insulation after a long stay in the cold reserve, are difficult to diagnose by traditional types of tests according to [1]. Therefore, in the main measures to prevent such violations in acts of investigating a technological violation, it may be suggested to include the oil circulation system before the transformer is put into operation, a re-physicochemical analysis of oil samples for moisture content, installation of a CR monitoring system on AT with its integration into the process control system [8, 11- 16].

Radial buckling in the HV winding of 250 MVA 500/110 kV autotransformer after a short circuit in the service is in Figure 5.1.

5.4. Damage of Power Transformer, Which Arose Because of External Short Current and Electrodynamic Action on the Busbar

The phase-to-phase short circuit on 10 kV busbar support insulators of 63 MVA/220 kV transformer installed on the traverse was occurred at 220 kV substation. Short circuit on 10 kV busbar was a result of a fall from above the foreign object, presumably wire, from the sockets located on the portal above the 10 kV bus bridge, with wind loads. The second reason is contamination of 10 kV insulators with emissions of cement dust into the atmosphere from a cement plant, on the balance sheet of which there is a 10 kV side. The porcelain insulation of the 10 kV bushings was destroyed from the electrodynamic effect of the short-circuit current with followed touching the 10 kV busbar to coolers, burning them, emitting and igniting the transformer oil (Figure 5.2) [17, 18].

10 kV circuit breakers of 1T was switch off on the 220 kV substation by the action of the differential protection of the transformer. 3CB circuit breaker of 1T has not disconnect. As a result of the failure of 3CB circuit breaker of 1T, and when the position of the sectional SCB-3 and 3CB of the second 2T switched on transformer, the short-circuit place was replenished within 3.5 seconds from the side of 2T. The fire was extinguished by joint efforts of the departmental fire brigade, the fire brigade of the Ministry for Emergency Situations, and the paramilitary fire brigade (only 12 cars).

The fire is qualified according as a short circuit in the electrical installation by National Standard RD 153-34.0-20.802-2002 [19-23].

Figure 5.2. Damage to the occurrence of a fire of 63 MVA/220 kV transformer (1T).

5.4.1. Causes of the Onset and Development of Technological Breakdowns

1. Fall of foreign object, presumably wire, from the nests located on the portal 1T above the bus bridge 10 kV under the influence of wind loads and contamination of 10 kV insulators with cement dust.

2. Failure to disconnect the 3CB 1T 10 kV circuit breaker due to the decrease in the voltage and the failure of the separation diodes in the capacitor bank.
3. Inconsistency of the circuit on the 10 kV side to the normal mode circuit (it does not correspond to the operational instructions for the power distribution system's relay protection (transformers 1T and 2T are included in parallel).

5.4.2. The Shortcomings of the Operation, Design, Construction, Manufacture, Installation of Equipment Were the Prerequisites of Technological Breakdowns or Complications of Liquidation

1. Availability of nests of birds on 1T portal and contamination of 10 kV insulators by cement dust.
2. Non-compliance with the cement plant circuit normal mode - 1T and 2T transformers worked in parallel on 10 kV side.
2. Absence of provisions on the relationship between the cement plant and the enterprise of the main electric networks.
3. Non-observance to comply with the timing of inspections and maintenance of relay protection devices of cement plant.
4. The lack of a scheme of normal operation and repair schemes for the duty personnel of the cement plant.
5. The absence of a 10 kV maximum current protection on the input cells 10 kV of 1T, 2T transformers and section circuit breaker SCB-1, SCB-2, SCB-3 of 220 kV substation.

5.4.3. The Main Measures to Prevent Such Technological Breakdowns

1. The extraordinary inspection of the 500 kV, 220 kV substation equipment conducted for the presence of nests and foreign objects on the portals of the switchgear. If it detected, take measures to

eliminate the comments, if necessary with the shutdown of equipment.

2. The joint test of the relay protection of input and sectional 10 kV circuit breaker conducted at 220 kV substations.
3. The long-term test schedule for the relay protection testing at substation 220 kV developed, taking into account the input 10 kV circuit breaker of 1T and 2T.
4. The position of relationships developed between an electrical network and the cement plant.
5. The insulation of 220 kV, 10 kV outdoor switchgear equipment purified and processed with hydrophobic pastes at 220 kV substation.
6. The scheme of normal and repair modes of 220 kV substation developed, agree with dispatch control and cement plant.
7. The 10 kV support insulation on 1T and 2T transformers bus bridges replaced with the installation of additional support insulators to improve mechanical stability against dynamic forces at short circuit.
8. A joint technical meeting with the management of the cement plant conducted to analyze the condition of the relay protection and automatic equipment of the 220 kV substation, their sufficiency and the development of a program to improve the reliability of the relay protection devices.

5.5. Internal Damages of Power Transformers as a Result of the Manufacture Defects

Insufficient electrodynamic withstand of tranformer windings when over rated, through-current short-circuiting currents along them can lead to a breakdown of insulation in the place of residual deformations and turn-to-turn short-circuits with severe consequences, especially for transformers with a service life of more than 25 years [23-31].

Separately, we give examples of damage due to manufacturing defects in the new transformer-reactor equipment:

- high gas evolution due to chromatography analysis results and subsequent damage of 167 MVA/500 kV/220 kV autotransformer (Figure 5.3),
- gas evolution and further damage of 60 MVA/500 kV reactor (Figure 4.1) [14-15, 26].

Figure 5.3. Internal turn-to-turn short-circuit and initial electrodynamic deformation of the LV winding of 167 MVA/500 kV/220 kV autotransformer.

The reason of 167 MVA/500 kV/220 kV autotransformer fault was inter-coil short-circuit with the formation of a source in the form of a gas bubble and 35 kV LV winding damage. Copper small balls, traces of combustion products, soot were found on the main rod in the exit zone of LV winding on the upper console. When inspecting the tap of LV winding in the zone adjacent to the winding, the darkening of the insulation of the upper turns of LV winding is visible. This internal turn-to-turn short-circuit and initial electrodynamic deformations of the LV winding of 167 MVA/500 kV/220 kV autotransformer were due to a defect in the winding wire - the presence of needle burrs with disturbance of paper insulation, i.e., factory defect and the development of intense partial discharges (PD) in a given local focus. After disassembling at the manufacturing plant and identifying the causes of damage, the autotransformer was repaired by factory warranty.

It is possible to form a focus with weakened insulation, which can exist for many years when operating transformer-reactor equipment with a significant service life at the deformation point of the winding. The development of the PD is gradually proceeding in it, leading ultimately to the breakdown of insulation and to turn-to-turn short-circuit [8, 26-39].

5.6. Internal Turn-to-Turn Short-Circuit in the Transformer Winding

As already noted, one of the causes of internal damages of windings is their electrodynamic instability at short circuits. This concerns first of all the transformers manufactured in the late 1960s and early 1970s, when the methods of electromagnetic calculations of windings for short-circuit withstand were imperfect and not confirmed by real electrodynamic tests.

Electrodynamic tests of power transformers with a power of over 63 MVA were carried out in Russia in the 1970s-1980s. and early 1990's at Shortcurrent Testing Laboratories - R&D Center of Federal Grid Company of Unified Energy System (Beskudnikovo), All-Russian ELectrotechnical Institute, Togliatti Power Testing Laboratory, etc. After 1994, short-circuits tests for withstand to short-circuit currents were carried out in Russia only for 16 MVA, 25 MVA, 40 and 63 MVA transformer capacities [6, 8, 21, 23, 26-29].

In subsequent years, the electrodynamic tests of power transformers with a power of over 63 MVA were carried out on the European stands of KEMA, CESI, or it was permitted to confirm the withstand to short-circuit currents only by calculation. This is not entirely correct, since the calculations do not always reflect the real state of the transformer windings in terms of withstand to short-circuit currents. All transformer design elements (winding, pressing system, and etc.) must be checked during short-circuit testing by real values of short-circuit currents. Transformer testing for short-circuit withstand is an instrument for reliability improvement of power transformer design.

In the late 1960s and early 1970s, the main manufacturing plants allowed some design errors in the design of the active part of power transformers, especially autotransformers (AT), which caused damage to transformer electrical equipment that had developed its normative resource. At numerous meetings, including in RAO UES of Russia, held in the early and mid-1990s, it was noted that 250 MVA/500 kV/110 kV autotransformer type, designed and manufactured in accordance with National Standard GOST 17 544-72, have increased damageability. As is known, since 1985 transformer manufacturers produce this autotransformer type with increased electrodynamic stability (for a pair of HV-MV sides, the voltage u_k is increased from 10.5% to 13%).

The most expensive electrical equipment operated at electric power stations and substations of grid enterprises – power transformers. Inside the transformer tank there are tens of tons of combustible substances – transformer oil, in case of fire, the fire can spread to nearby electrical equipment and structures. At the same time, the reliability of power supply to consumers is significantly reduced, the damage from its interruptions is increasing, the ecological situation in a number of located areas is deteriorating.

A typical example of this is turn-to-turn short-circuit in the regulating winding (RW), which occurred at August of 2003 due the destruction of the coil insulation paper, which led to the fire and failure of 250 MVA/500 kV/110 kV autotransformer at 500 kV substation of Siberian region, despite the impeccably worked stationary water fire extinguishing system. Differential and gas protection were worked, and autotransformer was swich off. Then there was an overlap of insulation of the 500 kV bushing of phase B. The hydraulic shock caused in this way destroyed 500 kV bushing and caused the release of combustion products of arc and oil through the bushing adapter. The burning oil from autotransformer expander, which did not have a shut-off valve on the pipeline of the gas relay, was mixed with the products of its decomposition from the arc (3000 oC) and RW copper. Then, the oil heated to 1400 ° C (near the place of turn-to-turn short-circuit of the winding) was ejected from the 500 kV bushing terminal of phase B. At the same time, superheated gaseous CxHy hydrocarbons and hydrogen ignited on contact with atmospheric oxygen.

After the explosion of the pressure tank of 500 kV bushing of phase A, the bushings and pressure tanks of other voltages and phases began to collapse with the release of burning oil outside the oil receiver. The autotransformer tank was deformed, there was a loss of tightness of the rubber gasket, there was an additional oil leakage from the tank, which ignited and intensified the fire.

External fire was localized only 5 hours after 250 MVA/500 kV/110 kV autotransformer swiching off, when oil leakage came to an end, and firemen were able to open repair hatches for connecting 500 kV bushings and to supply water inside autotransformer.

The flame emissions from the tank stopped, but the combustion inside the active part of autotransformer continued for another two days until the oil and insullation were completely burnt out. The transformer was completely destroyed. The cost of a new autotransformer is about 2 million US dollars [6, 8, 21, 23, 26-29].

The appearance of 250 MVA/500 kV/110 kV autotransformer after the fire and destruction of bushings is shown in Figure 5.4, its windings located from the magnetic core in the following order: LV, RW, GW (general - 110 kV), SW (serial – 500 kV), – in Figure 5.5 (the focal point of turn-to-turn short-circuit is shown by an arrow). The accident at 500 kV substation confirms the statistics of increased damage to these autotransformer type due to the dynamically unstable single-layer RW, which also has insufficient cooling.

Another case of unplanned swiching off due to differential and gas relay protection took place with 125 MVA/220 kV/110 kV autotransformer at 220 kV substation of Central region. Through the membrane of the exhaust pipe, there was a slight ejection of oil, a combustible gas was detected in the gas relay, no apparent deformations of the autotransformer tank were observed.

Diagnostics using the low-voltage pulse method (LVI) was carried out in order to control the geometry of autotransformer windings and to determine its maintainability. Significant differences in the oscillograms of the LVI of the LV, RW and GW windings of autotransformer at 220 kV substation illustrating the sharp difference in phase C, especially noticeable in differential differences involving phase C (a – c and b – c), confirm the presence significant electrodynamic deformations and turn-to-turn short-circuit in the windings.

Figure 5.4. The appearance of 250 MVA/500 kV/110 kV autotransformer after the fire and destruction of bushings.

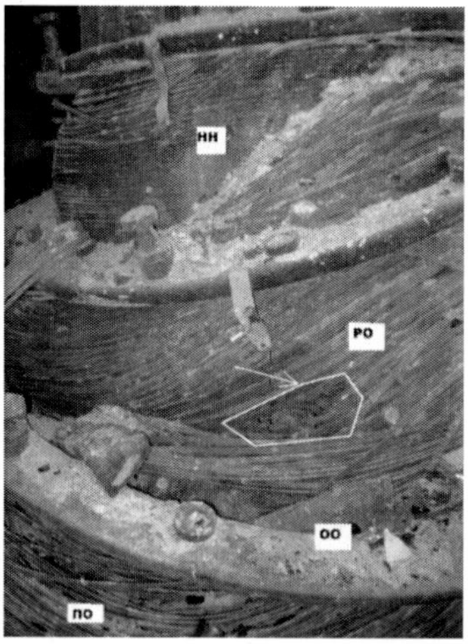

Figure 5.5. Windings of 250 MVA/500 kV/110 kV autotransformer after the fire located from the magnetic core in the following order: LV, RW, GW (general - 110 kV), SW (serial – 500 kV), the focal point of turn-to-turn short-circuit is shown by an arrow.

Serious deformations and turn-to-turn internal short-circuit were detected in MV 110 kV winding, regulating winding and LV winding by LVI-testing, short-circuit inductive reactance measurements and iron core losses methods. The short-circuit impedance differences of C phase from A and B phases are Δ Zs-c= -11.6% in MV-LV winding regime, and Δ Zs-c= -7% in HV-LV winding regime.

The 125 MVA/220 kV/110 kV autotransfomator at 220 kV substation of Central region was recognized as non-maintainability and subject to disposal[8, 21, 29-37].

Operational personnel should clearly represent that the unplanned outages of the critical power transformers (manufactured in the 1960s and 1970s) associated with the passage of short-circuit currents through their windings and the operation of gas and differential protection. This makes it necessary to conduct a diagnostic survey of the active part of the transformer.

It is not possible to limit oneself only to measuring the parameters R of insulation, tgδ, DC ohmic resistance of windings - Rom, no-load losses, etc. It is also necessary to determine the degree of polymerization of the paper insulation wire, to measure short-circuit impedance Zk and to perform FRA or LVI diagnostics, since it is more sensitive to any changes in the geometry of the windings (local small interturn, inter-coil, interfacial capacitances and inductances). Measuring only the short-circuit impedance Zk is effective only with significant changes in the inductance of the windings, if residual deformations cover a large number of turns and a violation of dielectric strength of winding insulation can occur [39-47].

All responsible 220 - 750 kV power transformers and autotransfomator, manufactured in the 1960s and 1970s, must undergo a comprehensive survey using methods specified in the normative document National Standard “Volumes and standards for testing electrical equipment”, including evaluation by degree of polymerization. Samples of the paper of the coil insulation of the upper coils should be selected during planned capital maintaines to determine the remaining life of the winding cellulose destruction. Degree of destruction of cellulose is very important when deciding whether to replace a transformer with a new one, depending on the costs of repairs and the condition of its structural parts.

The places of seizure of paper must be coordinated with technical inspection and with the leading organization in this direction – Research Institutes [8, 26, 48].

In addition, it is necessary to take into account all cases of unplanned outages of transformers and autotransfomator for the past years, and also, if necessary, carry out FRA or LVI diagnostics to detect residual deformations after the passage of short-circuit currents through the transformers windings. Thus, it is possible to avoid the accidents mentioned in the examples of 500 kV and 220 kV substations. After a comprehensive survey, a conclusion should be drawn about the possibility of further operation of the transformer and the frequency of repeated measurements is established if the transformer remains in operation.

CONCLUSION

1. All responsible 220-750 kV power transformers with a capasity of over 125 MVA, manufactured in the 1960s and 1970s, must undergo a comprehensive survey using methods specified in the normative documents, National Standards. In addition, in the transformer under test, the degree of polymerization of the coil paper insulation (with mandatory calculation of the remaining life) should be evaluated, especially in transformers with a polymerization degree of 300 to 450 units in the risk zone.
2. It is necessary to identify all cases of unplanned outages for the past years, assess their danger and take them into account in the future. If necessary, in addition, FRA or LVI diagnostics should be carried out in order to detect residual deformations after the passage of short-circuit currents through the windings of transformers, which in most cases will help to avoid internal turn-to-turn short-circuit and serious accidents with fires.

3. The conclusion should be made about the possibility of further operation of the transformer after a comprehensive survey, the frequency of repeated measurements and the load capacity of the transformer is established, ensuring its trouble-free operation until replacement or capital repairs, if it remains in operation.
4. It is necessary to introduce systems that prevent the ignition of transformers and reduce environmental damage. These are, for example, the protection systems of the transformer tank and internal fire extinguishing nitrogen system of SERGI company, which should be equipped with all the newly ordered transformers.
5. It is recommended that the National Standard and normative document of volumes and standards for testing electrical equipment be supplemented with the following indications on the types of measurement when commissioning new transformers: phase-by-phase measurement of the short-circuit impedance (Zk) in all modes, in the nominal and two extreme positions of the on-load tap-changer; removal of FRA or LVI normograms, determination of primary data on PD, primary thermograms of thermal infra-red control, data on pressing of windings.
6. All new transformers should be equipped with monitoring systems (installed and now at many sumstations) for the following parameters: current, voltage, oil temperature, chromatographic data of combustible gases dissolved in the transformer oil, PD intensivity, tg δ, winding geometry (Zk or leakage inductance), etc., and fire extinguishing systems on the principle of filling the volume of the tank, for example with compressed nitrogen, preventing the ignition of the transformer for any damage.

ACKNOWLEDGMENTS

To Mr. Alexander Rubtsov from Irkutsk, Mr. Nikolaj M. Alexandrov, post-graduate student, JSC “Dinamika”, Department of Technical Inspection of Federal Grid Company of UES, Russia, Cooperation of

Universities and Innovation Development, Doctoral School project "Complex diagnostic modeling of technical parameters of power transformer-reactor electrical equipment condition", 2009 has made publishing of this article possible.

REFERENCES

[1] IEC 60076-1, CEI 60076-1. *Power transformers – Part 1: General.* 2011, 154 p.

[2] IEC 60076-2, *Power transformers – Part 2: Temperature rise for liquid-immersed transformers*. 2011, 100 p.

[3] IEC 60076-3, CEI 60076-3. *Power transformers – Part 3: Insulation levels, dielectric tests and external clearances in air*. 2013, 124 p.

[4] IEC 60076-5. *Power Transformers – Part 5: Ability to withstand short circuit.* 2006, 80 p.

[5] The short circuit performance of power transformers. *CIGRE. Working Group* 12.19. 2002, 43 p.

[6] Bertagnolli, G., (1998). Short-Circuit Duty of Power Transformers." *ABB Transformatori*, Legnano (Milano) – Italy, 1997.

[7] Wang, M.Y. (2001). 1995-1999 Fault Statistics & Analysis of All Transformer Typesin China. *Electrical Equipment*, 49(1),p.11-12.

[8] Khrennikov, A.Yu. *Diagnostics of Electrical Equipment Faults and Power Overhead Transmission Line Condition by Monitoring Systems (Smart Grid): Short-Circuit Testing of Power Transformers,* Nova Publishers, New York, 2016, USA, p. 165. https://www.nova publishers.com/catalog/product_info.php?products_id=56561

[9] Shige, W.. *Fault Analysis and Technical Improvements*, China Power Press, Beijing, (2004).

[10] Ying, X., Shixian, X. *AC Power System Over Voltage Protection and Insulation Coordination,* China Power Press, Beijing, (2006).

[11] Fogelberg, T., Girgis, R.S., (1994). ABB power transformers - a result of merging different technologies with prospects for significant future

advancements", *International Simposium "Electrotechnics - 2010"*, Moscow, vol.1, 56-59.

[12] Balakrishnan, V., (1989). A study of short-circuits in large power transformers, *Electrical India,* Vol. XXIX, № 4, 87-91.

[13] McNutt, W.J., Johnson, W.M., Nelson, R.A., Ayers, R.E., (1970). Power transformer short-circuit strength - requirements, design, and demonstration. *IEEE Trans. Power Appar. and Syst.,* 89, № 8.

[14] Malewski, R., Khrennikov, A.Yu., Shlegel, O.A., Dolgopolov, A.G. (1995). *Monitoring of Winding Displacements in HV Transformers in Service,* Reports of CIGRE Working Group 33.03. Italy, Padua.

[15] Khrennikov, A.Yu., (1999). *Short-circuit performance of power transformers. Test experience at Samaraenergo Co and at Power Testing Station in Togliatti, including fault diagnostics,* Reports of CIGRE Study Commitee 12 Transformers. Hungary, Budapest.

[16] Khrennikov, A.Yu., (2000). *Power transformer's fault diagnostics at Samaraenergo Co, including FRA/LVI method,* Reports from School of Math. and System Engineering, Vaxjo University, Sweden, № 43, ISSN 1400-1942.

[17] Kulkarni, S.V., Khaparde, S.A. *Transformer Engineering. Design and Practice*//Marcel Dekker, INC. NY, Basel, 2004, 478 p.

[18] Del Vecchio, R.M., Poulin, B., Feghali, P.T., Shah, D.M., Ahuja, R. *Transformer Design Principles With Applications to Core-Form Power Transformers*//Second Edition. CRC Press, Taylor & Francis Group, NY, 2010, 602 p.

[19] Shintemirov, A. *Intelligent Modelling and Condition Assessment of Power Transformers*. LAP Lambert Academic Publishing, 2013, 308 p.

[20] He, Zh.Q., Gao,W.L. (2010).Analysis on restrictive measures against short-circuit current of 220kV Chongqing power grid. *Central China ElectricPower*, 23(2),p.31-34.31.

[21] Khrennikov, A.Yu., Shakaryan, Y.G., Dementyev, Y.A., (2013). Shortcurrent Testing Laboratories. Short-Circuit Performance of Power Transformers, Transformer Testing Experience. *International*

Journal of Automation and Control Engineering (IJACE), Volume 2, Issue 3, 120-127. http://www.seipub.org/ijace/AllIssues.aspx.

[22] Kravchenko, E.V.A, Ivleva, D.Yu. Thermal analysis of power transformer. *MATEC Web of Conferences* Volume 23, 2015, p. 4.

[23] Janssen, A.L.J., te Paske, L.H., Smeets, R.P.P., Leufkens, P.P., Fogelberg, T., (2009). YEARS TEST Experience with short-circuit withstand capability of large power transformers. *CIGRÉ 2009, 6th Southern Africa Regional Conference,* Paper P501.

[24] Abdallah, J., (2009). Using the Frequency Response Analysis (FRA) In Transformers Internal Fault Detection. *WSEAS Transactions on Power Systems,* Issue 9, Volume 4, 297-306.

[25] Khrennikov, A.Yu., (2012). New "intellectual networks" (Smart Grid) for detecting electrical equipment faults, defects and weaknesses. *Smart Grid and Renewable Energy,* Vol. 3, No 3, 159-164. http://www.scirp.org/journal/sgre.

[26] Khrennikov, A.Yu., (20130. Fault Detection of Electrical Equipment. Diagnostic Methods. *International Journal of Automation and Control Engineering (IJACE)*, Vol.2, No.1, 19-27. http://www.seipub.org/ijace/Archive.aspx.

[27] Khrennikov, A. Yu., (2013). *Smart Grid technologies for Detecting Electrical Equipment Faults, Defects and Weaknesses.* Workshop on Mathematical Modelling of Wave Phenomena with applications in the power industry, Linnaeus University, Växjö, 23-24. http://www.lnu.se/mmwp.

[28] Lapworth, J.A., (2001). Recent Developments Relating to the Detection of Winding Movement in Transformers by Frequency Response Analysis (FRA). *Sixty-Seventh Annual International Conference of Doble Clients.*

[29] Lech, W., and Tyminski, L., (1966). Detecting transformer winding damage - the low voltage impulse method. *Electrical Review,* № 18.

[30] IEEE Std C57.12.90-1993, Part II: *IEEE Guide for Short-Circuit Testing of Distribution and Power Transformers.*

[31] Janssen, A.J.L., te Paske, L.H., (2000). Short-Circuit testing experience with large power transformers, *CIGRE SC12 Session,* report 12–105.

[32] Janssen, A.L.J., te Paske, L.H., Smeets, R.P.P., (2002). Short-circuit Tests on Large Transformers: Destructive or not? *TrafoTech,* Mumbai, India, 126-33.

[33] Janssen, A.L.J., te Paske, L.H., van der Linden, W.A., Smeets, R.P.P., (2001). Magnetic Saturation at Short-Circuit Tests on Power Transformers, *IEEE PES Summer Meeting,* paper 72.03.

[34] IEC Committee Draft 14/465/CD Annex B, (2003). *Theoretical Evaluation of the Ability to Withstand the Dynamic Effects of Short Circuit.*

[35] CIGRE WG12.05, (1983). An International Survey on Failures in Large Power Transformers in Service, *Electra* 88, 21–37.

[36] Bertagnolli, G., (1999). Results of short-circuit performance of transformers, CIGRE Transformer Colloquium, Budapest, Short-circuit performance*; Tests and Failures, report 2, CIGRE TC 12.*

[37] Bartley, W.H. *Transformer Asset Management.* http://www.hsb.com/TheLocomotive/TransformerAssetManagement.aspx.

[38] Khrennikov, A.Yu., (2009). *Complex diagnostic modeling of technical parameters of power transformer - reactor electrical equipment condition.* PhD Thesis, Samara State University (in Russian), 405.

[39] Perkins, M., L. Pettersson, N.L. Fantana, T.V. Oommen, S. Jordan. Transformer Life Assessment Tools with Special Application to Nuclear Power Station Generator Transformers. *IEEE Transformer Committee Meeting,* November 1999, Monterrey Mexico.

[40] Woodcock, D. J., Weidmann Systems International, Inc. “Condition-Based Loading of Power Transformers” *Sixty-Ninth Annual International Doble Client Conference*, 2002.

[41] Lindgren, S. & Ward, B., EPRI, “Management of Aging Power Transformers” *CEA Electricity Conference,* 1999.

[42] Aubin, J., Bourgault, A., Rajotte, C., Gervais, P., Profitability Assessment of Transformer On-Line Monitoring and Periodic

Monitoring, *EPRI Substation Equipment Diagnostics Conference*, February 2002.

[43] Carlson, A., Fuhr, J., Schemel, G., Wegscheider, F. Testing of Power Transformers. *Routine tests, Type tests and Special tests*, 1st Edition published by PRO PRINT for ABB. https://ru.scribd.com/doc/100174040/Testing-of-Power-Transformers.

[44] Farhadzadeh, E.M., Muradaliyev, A.Z., Rafiyeva, T.K, Ismayilova, S.M. The computer analysis of faultlessness transformers of power supply systems. *RT&A* № 01 (20), (Vol.2) March 2011.

[45] Cujbă, T., Cernomazu, D. Contributions to the development of devices to increase safety in operation of the on-load–tap-changers of the power transformers. *U.P.B. Sci. Bull.*, Series C, Vol. 73, Iss. 2, 2011 ISSN 1454.

[46] Wang, M.Y. (2006). 2001-2005 Fault Statistics & Analysis of All Transformer Typesin China. *Electrical Devices,* 37(2), p.4-13.

[47] Zhang, K.L. (2001). Improve the ability of large power transformer short-circuit resistance measures. *Fujian Dianli yu diangong*, 1(3), p.8.

[48] Lvov, M.Yu., Komarov, V.B., Lvov, Yu.N., etc. Aging of cellulose insulation of power windings transformers in the process of operation. *Electrical Stations*. 2004, № 10 (in Russian).

Chapter 6

DIAGNOSTIC AND MATHEMATICAL MODEL OF THE WINDINGS OF THE HIGH-VOLTAGE TRANSFORMERS AND REACTORS: ELECTROMAGNETIC FORCES

A. Yu. Khrennikov[*], *PhD*

ABSTRACT

The chapter shows diagnostic and mathematical models of the windings of the high-voltage transformers and reactors. Electromagnetic forces acting on the windings of a two-winding transformer are also considered. The case of finding the current-flowing conductors of the transformer in a magnetic leakage field the electrodynamic forces act on them. Mathematical models of winding radial and axial deformations are received.

[*] Corresponding Author's E-mail: ak2390@inbox.ru.

Keywords: deformation of the winding, electromagnetic forces, diagnostic models, mechanical state, flux density, electromagnetic forces, hoop buckling, radial and axial forces

6.1. Introduction

The causes of the deformation of the windings are electromagnetic forces that arise and are applied to a conductor located in a magnetic field. The electromagnetic force acting on a winding is directly proportional to the winding current and the flux density. In addition, this force is also proportional to the square of the winding current, since the leakage field can be expressed in terms of the winding current. All this indicates that during short circuits the forces in the winding are several times higher than with the flow of rated currents [1].

Diagnostic models examined make it possible to understand and to describe the nature of processes inside the transformer windings under the action of external operational physical actions and to detect defects in the active part at the early stage of their appearance.

6.2. Diagnostic Model of the Electrodynamic Stability (Mechanical State) of the Transformer and Reactor Electrical Equipment (TREE) Windings

6.2.1. Electromagnetic Forces Acting on the Windings of a Two-Winding Transformer

The diagnostic model of the electrodynamic stability (mechanical state) of the transformer and reactor electrical equipment (TREE) windings to the short-circuit currents makes it possible to forecast the residual the service life of the materials of the construction of transformer, and also the processes, proceeding inside the transformer.

Current-flowing conductors of the transformer windings are in a magnetic leakage field, and electrodynamic forces act on them. These forces create mechanical stresses in the material of the transformer windings and are partially transmitted to other elements of its construction. At nominal currents, the forces are small, but with short circuits, they greatly increase and can cause a transformer failure. The transformer winding consists of copper or aluminum coils separated by insulating spacers. The dynamic characteristics of such a system change during a short circuit, since the elasticity of the insulation pads depends on the degree of their compression, i.e., on the magnitude of the acting forces. In turn, the forces do not remain constant, but change according to complex laws. The winding has several natural frequencies. In the event that one of the frequencies of the electrodynamic forces coincides with the frequency of natural oscillations, a resonance occurs and, as a result, an increase in pressure above the static one. The unsteady process in the winding is the superposition of two transients: electromagnetic and mechanical. When calculating the mechanical strength of a transformer, until recently, the windings were considered as stationary, and the forces acting on them - as constant, corresponding to the maximum current flowing through them.

In other words, the dynamic task was replaced by a static one. In recent years, scientific works have appeared that indicate the need to consider the dynamics of the process in all its complexity [2-8].

The electromagnetic force acting on an element of a current in a magnetic field can be determined according to the law of Bio-Savart in differential form:

$$df = [Bj]dv \tag{6.1}$$

where df is the force vector, acting on the conductor element volume dv, located in a field with induction B at a current density j. The vector product on the right side of the equation shows that the force is perpendicular to the direction of induction and the direction of current density. Consequently, the axial field creates radial forces, and the radial (transverse) - axial. The directions of force, induction and current density are related by the left hand rule.

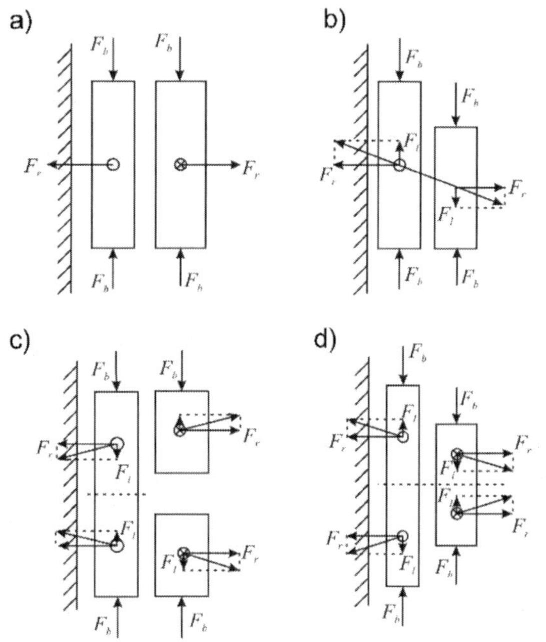

Figure 6.1. Electromagnetic forces acting on the windings of a two-winding transformer.

The expression for the force acting on the unit length of each of the two parallel wires flowing around the currents i1 and i2 is used to roughly estimate the direction of the forces in the transformer windings:

$$f = \frac{\mu_0 i_1 i_2}{4\pi d} \tag{6.2}$$

where μ_0 is the magnetic permeability of the medium where the wires are located, and d is the distance between them [9-14].

Wires with current are attracted at the same directions of the currents and repel each other. All the turns of one winding are flowed around by currents of the same direction, therefore they are attracted to each other. On the contrary, the currents in the primary and short-circuited secondary windings are phase shifted by π, so these windings are repelled.

When using formula (6.2), it is natural to divide all forces acting on the transformer windings into internal ones, i.e., interaction forces between elements of one winding and external ones, i.e., interaction forces between different windings.

The simplest cases of the windings of a two-winding transformer are indicated; the directions of forces are shown in Figure 6.1.

In Figure 6.1a shows two windings of equal height with a uniform distribution of the magnetically moving force. External forces can be approximately considered to be directed along a straight line connecting the centers of gravity of the winding sections. In this case, they are purely radial and are denoted by *Fr*. These forces stretch the external winding, strive to break it, and compress the internal one (Figure 6.2). Own forces in figrure 39a are axial, seeking to reduce the height of the windings.

Consider the magnetic leakage field of scattering of a transformer with concentric windings with a uniform distribution of the magnetically moving force along their height.

In the middle part of the winding, the field lines are parallel to the axis of the rod, and they are bent at the ends of the windings, i.e., in addition to the axial, the radial component of the field appears. It also shows that both radial and axial electromagnetic forces act on the windings. If we calculate the axial and transverse components of the induction of the magnetic leakage field, we obtain the diagrams of the induction shown for one of the windings in Figure 6.3. Induction of the axial field *BY* causing radial forces, in the middle part of the winding has a constant value, decreasing somewhat towards its edges. The induction of the transverse field *BX*, which creates axial forces, has the greatest value at the ends of the winding and passes through zero in the middle of its height, changing sign. Own axial forces bend the winding coil in the vertical direction and compress the gaskets between the coils. The greatest bending forces occur in coils located at the ends of the winding, since the force is proportional to the induction.

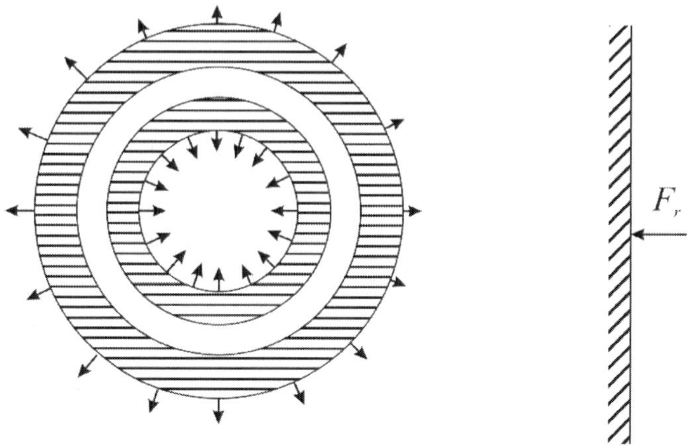

Figure 6.2. Radial forces acting on the windings double winding transformer.

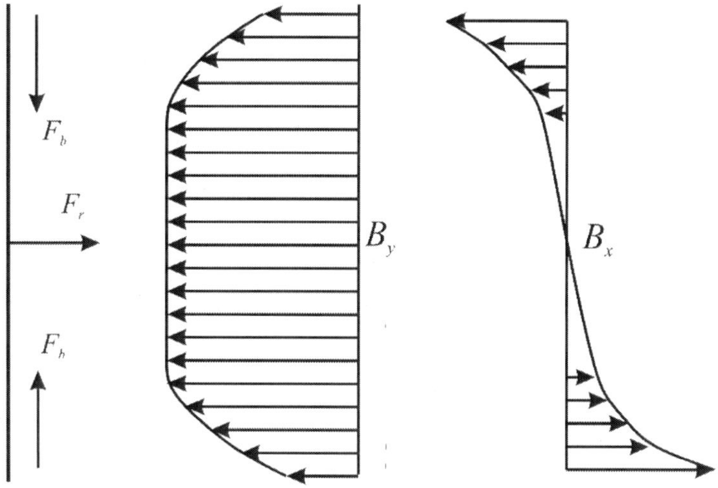

Figure 6.3. Distribution of longitudinal and transverse *Bx* induction and transformer with windings of equal height.

However, the greatest compressive forces are experienced by their gaskets located in the middle of the winding height, since the sum of all forces acting on all winding coils is transferred to these gaskets.

In Figure 6.1b shows the case of unilateral shortening of one of the windings. The difference between this case and the previous one is that the external forces directed in the figure along a straight line connecting the centers of the winding sections already have not only the radial components *Fr*, but also axial *Fl*. External axial forces appear in the windings in cases where the windings have an unequal height (mainly due to adjusting spurs) or an uneven distribution of magnetically moving force. These forces are always directed in such a way as to increase the asymmetry that created them. External axial forces more often than other forces are the cause of accidents, therefore, when designing transformers, they always try to reduce the asymmetry in the distribution of magnetically moving force of windings. The external axial forces *Fl*, as well as the own axial forces, bend the coil in the vertical direction and compress the spacers between the coils. In addition, they are partially or completely transferred to the magnetic yoke of the core, seeking to tear them from the rods.

In Figure 6.1c the case is given when the de-energized part of one of the windings is in the middle of its height, due to the placement of the regulated zone here. Such an arrangement of the regulated zone is very often used in high-power transformers, since it reduces the transverse field and the associated external axial forces. In the figure, forces are conventionally directed along straight lines connecting the centers of gravity of the sections of the nearest halves of the windings. This is done because we are interested here only in the directions of action of the forces. External axial forces of the internal winding in Figure 6.1c are directed to the same side as the internal axial forces, so they do not create pressure on the yoke, but increase the pressure on the gaskets in the middle of the winding height, which then becomes equal to $Fb + Fl$. External axial forces in the outer winding are opposed to internal axial, therefore the pressure on each of the magnetic yokes is equal to the difference $Fl - Fb$, provided that $Fl > Fb$, or equal to zero, if $Fl < Fb$. The most compressed at the same time are the gaskets on the ends of the windings.

In Figure 6.1d shows the case when the windings have different height, but are located symmetrically relative to each other. The direction of the forces is clear from the figure.

6.2.2. The Loss of the Radial Stability (Radial Buckling) of Winding. Calculation of the Forces Acting on the Concentric Windings of a Two-Winding Transformer

Electromagnetic Forces

The forces acting on the transformer windings can be calculated using either the law of Bio-Savart or the Lagrange theorem.

In the first case, the force acting on the winding or part of it, can be calculated by integrating equation (6.1):

$$\boldsymbol{F} = \int_V [Bj]dV\mathrm{F} \quad (6.3)$$

If B and j are perpendicular to each other and constant throughout the volume, then

$$F = Bli, \quad (6.4)$$

where l - is the length of the coil or winding wire, a I - is the current in it.

The Lagrange theorem as applied to an electromagnetic field can be formulated as follows: in a system of circuits with a current in a magnetic field, an electromagnetic force that tends to change a given coordinate of the system is equal to the derivative of the energy of the magnetic field on this coordinate under the assumption that the currents in the circuits maintained constant. Considering the transformer windings as separate circuits, we get

$$f_g = \frac{\partial W_M}{\partial g} \quad (6.5)$$

where f_g - is the force acting in the direction of the coordinate g; WM- energy of the magnetic leakage field of the transformer windings.

$$W_M = \frac{1}{2}\mu_0 \int_V B_\sigma^2 dV \quad (6.6)$$

Or

$$W_M = \frac{1}{2} L_K i^2 \tag{6.7}$$

where L_K - is the inductance of the transformer short circuit. In this way,

$$f_g = \frac{\partial W_M}{\partial g} = \frac{1}{\mu_0} \int_V B_\sigma \frac{\partial B_\sigma}{\partial g} dV \tag{6.8}$$

Both of these methods give fairly accurate results if the picture of the distribution of the magnetic leakage field is correctly taken into account.

When determining the forces according to the Lagrange theorem, the use of energy expression through field induction (6.6) presents great difficulties, therefore, usually use the expression of energy through winding leakage inductance (6.7) and calculate only the total forces acting on the windings. The calculation of the leakage inductance involves a number of assumptions. For example, it is assumed that the distribution pattern of the induction of the leakage field is similar to the distribution diagram of the magnetically moving force. When calculating the leakage inductance from a longitudinal field, this does not introduce a significant error, and therefore the calculation of the total radial forces acting on the windings is quite accurate. However, when calculating the total axial forces associated with the transverse field, the error is greater and it is necessary to introduce correction factors, as shown below.

When calculating forces using the Bio-Savart theorem, a more accurate computational pattern of the stray field can be used, which takes into account the influence of rods, magnetic yokes, curvature of windings, etc., or the field pattern obtained on the transformer model by experiment. This provides greater accuracy. The Bio-Savart theorem makes it possible to calculate both the forces acting on individual turns or coils, and the total force acting on a winding or a part of it.

We first show the application of the Lagrange theorem.

The radial forces *Fr* acting on the concentric windings of a two-winding transformer tend to increase the distance between the windings $\delta 12$ (Figure

6.4). Therefore, in accordance with equation (6.8), it is necessary to take the derivative with respect to *δ12.* Substituting, instead of current *i*, the current amplitude of a sudden short circuit *Imax*, we obtain

$$F_r = \frac{1}{2} I^2_{\text{макс}} \frac{\partial L_K}{\partial \delta_{12}} \tag{6.9}$$

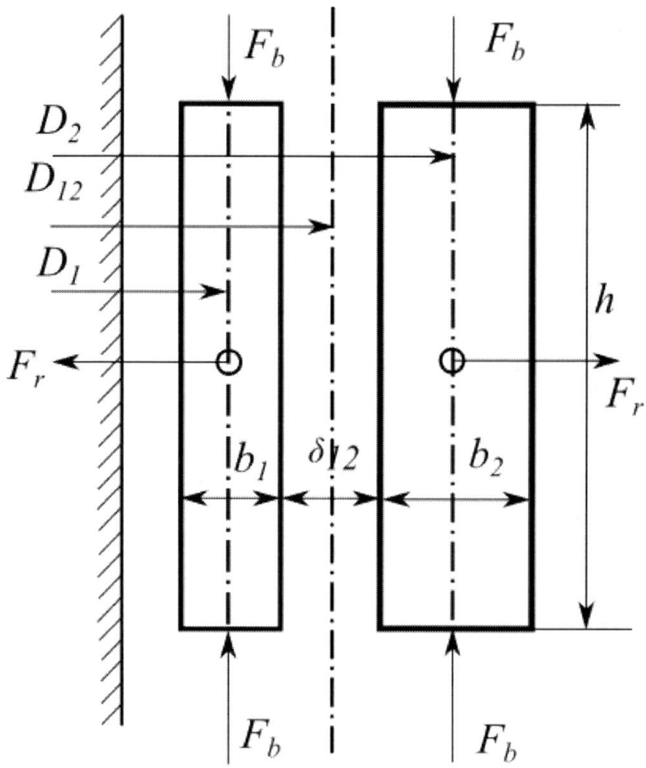

Figure 6.4. Calculation of the radial forces acting on the windings of the transformer.

Short circuit inductance:

$$L_K = \frac{\mu_0 \omega^2 \rho_1 \pi D_{12}(\delta_{12} + \frac{b_1+b_2}{3})}{h} \tag{6.10}$$

Then the radial force in newtons:

$$F_r = \frac{1}{2} I_{max}^2 \frac{\partial L_k}{\partial \delta_{12}} = \frac{\mu_0 (I_{\text{макс}} \omega)^2 \rho_1 \pi D_{12}}{2h} \qquad (6.11)$$

where μ_0 - is magnetic conductivity;

Imax - short circuit current amplitude;

ω - the number of turns of the winding through which short-circuit current flows;

ρ_l - the coefficient of reduction of the ideal leakage field to the real (Rogowski coefficient);

D12 - the distance of the tank wall to the middle of the gap between the windings (Figure 6.5);

h - winding height.

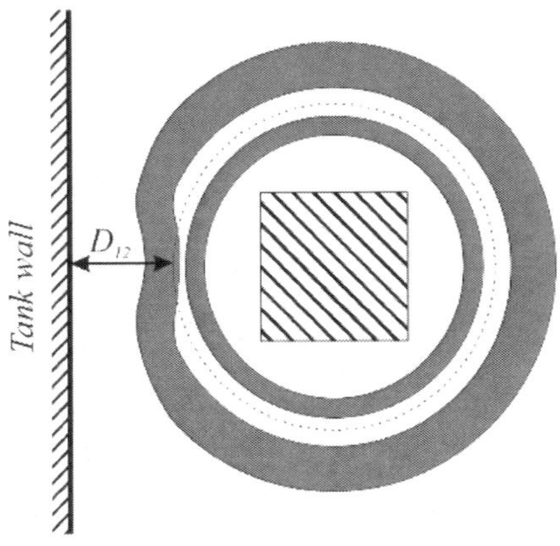

Figure 6.5. The distance D_{12} of the tank wall to the middle of the gap between the windings.

Equation for deformations of LV winding of 250MVA/220 kV transformer after short-circuit testing of the phase “C” with the loss of the radial stability (radial buckling) of winding is given [3, 15-23].

Thus, for instance, the diagnostic model of the electrodynamic stability (mechanical state) of the TREE windings [by 9-10] to the short-circuit currents makes it possible to forecast the residual the service life of the materials of the construction of transformer, and also the processes, proceeding inside the transformer. As a result can be introduced changes in construction TREE, which increase its reliability and which improve the indices of fitness for work. The mechanical state (deformations) of the TREE windings also can be estimated with the aid of the use of a method of Low-Voltage Impulses (LVI) [by 6, 9-11] (differential estimation, the penetration of the diagnosing impulse inside the winding) and on the basis of measurement data to the short-circuit impedance Zk (integral estimation, the collection of diagnostic information from the external outputs of transformer - parameters U, I and other) [11]. Deformations of LV winding of 250MVA/220 kV transformer with the loss of the radial stability (radial buckling) of winding are in the Figure 6.6 [21].

Figure 6.6. Deformations of LV winding of 250MVA/220 kV transformer after short-circuit testing of on the phase "C" with the loss of the radial stability (radial buckling) of winding.

Mathematical Model of Winding Radial Buckling

Next, a mathematical model is given for the radial deformation (example in the Figure 6.6) of one turn of a transformer winding in the form of an equation, which is more accurate than the previously presented models. Due to the use of trigonometric functions when describing the shape and volume of deformation and smaller error, this model more realistically reflects the processes occurring in the windings of power transformers in the case of short circuit.

Assumptions adopted in the preparation of the model:

1. The maximum size of the bulge of transformer winding is taken equal to the size of the concavity: a;
2. The convexity range is assumed to be equal to the concavity range: $\angle \Phi/2$;

As an equation describing the concave-convex deformation of the current-conducting coil turn, the function ε (ϕ), described in polar coordinates, is chosen (6.12) (Figure 6.7):

$$\varepsilon(\phi) = a \cdot \sin(2 \cdot \pi \cdot \phi/\Phi), \text{ with } 0 \leq \phi < \Phi, \quad (6.12)$$

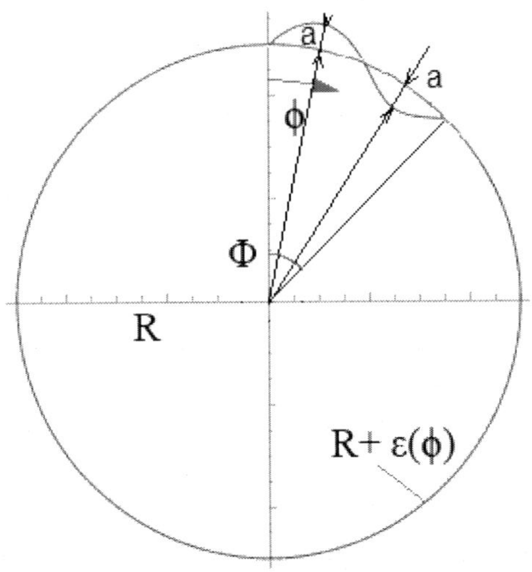

Figure 6.7. Top view of the cylindrical winding of a power transformer with a concave-convex deformation after short-circuit with radial buckling.

where a - maximum deformation, the maximum deviation of the distorted section from the ideal state;

$\angle \Phi$ - deformation range, the central angle within which the deformation is observed;

ϕ - is the function argument, the angle that is set clockwise from the vertical semiaxis.

The equation of radial deformation:

$$e(\phi) = a*\sin(2*\pi*\phi/\Phi), \text{ with } 0 \leq \phi < \Phi. \quad (6.13)$$

The equation describing the shape of the turn of coil:

$$(\phi) = R + e(\phi) = R + a*\sin(2*\pi*\phi/\Phi), \text{ with } 0 \leq \phi < \Phi; \quad (6.14)$$

R - winding radius, with $\Phi \leq \phi < 2*\pi$

Substituting in the formula (6.11) the expression for the radial concave-convex deformation (6.12), for reasons of changing the position of the middle of the gap between the windings, we have for radial force:

$$F_r = \frac{\mu_0(I_{max}\omega)^2\rho_1\pi(D_{12}+a\cdot\sin(2\pi\phi/\Phi)}{2h} \quad (6.15)$$

where μ_0 - is magnetic conductivity;

Imax - short circuit current amplitude;

ω - the number of turns of the winding through which short-circuit current flows;

ρ_1 - the coefficient of reduction of the ideal leakage field to the real (Rogowski coefficient);

D12 - the distance of the tank wall to the middle of the gap between the windings;

h - winding height.

Φ - deformation range, the central angle within which the deformation is observed;

ϕ - is the function argument, the angle that is set clockwise from the vertical semiaxis.

The expression (6.15) for the radial force acting on the transformer windings in the event of a radial hoop buckling in case of a short circuit has greater accuracy than other expressions.

This expression (6.15) for the radial force more accurately reflects the processes occurring in the windings of power transformers in case of short circuit and the occurrence of radial hoop buckling, due to the use of trigonometric functions in describing the shape and volume of deformation, Φ - deformation range, the central angle within which the deformation is observed, and h - winding height [24-30].

6.2.3. Axial Deformations

An example of the axial deformation of the LV winding of 250MVA/220 kV transformer, i.e., the movement of the winding in the axial (vertical) direction under the action of electrodynamic forces during the flow of short-circuit currents (Figure 6.8) [31-33].

Figure 6.8. Photo of phase "B" of the low voltage winding of 250MVA/220 kV transformer, illustrating the loss of the axial stability of the winding, i.e., the movement of the winding in the axial (vertical) direction under the action of electrodynamic forces during the flow of short-circuit currents.

Axial Forces

Similarly (6.8) – (6.11), own axial forces *Fb* (in newtons) acting on the windings tend to reduce the height of the windings, therefore, the derivative of the magnetic field energy must be taken along the height of the winding h, i.e.

$$F_{b(2)} = \frac{\partial W_M}{\partial h} = \frac{1}{2} I^2_{\text{макс}} \frac{\partial L_k}{\partial h} = -\frac{\mu_0 (I_{\text{макс}} \omega)^2 \rho_1 \pi D_{12} \delta'}{2h^2} \quad (6.16)$$

where *Fb(2)* - own axial force acting on both windings.

If we consider the axial deformation as a change in the height of the winding, then the axial deformation formula will look like:

$$\Delta = h_{HV} - h_{LV}. \quad (6.17)$$

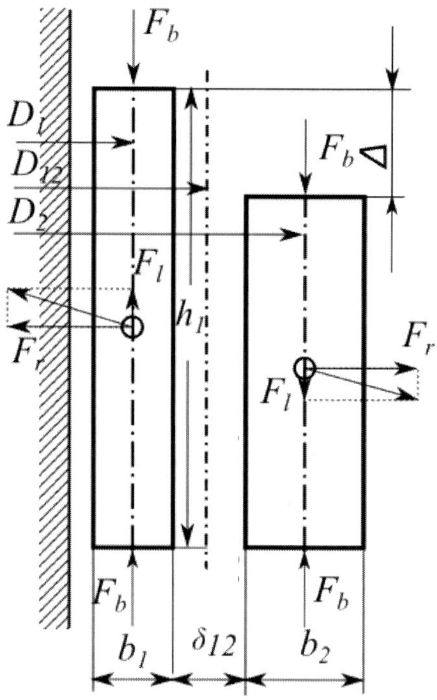

Figure 6.9. Calculation of internal axial forces acting on the windings of the transformer.

It follows that:

$$h_{HV} = h_{LV} + \Delta \quad (6.18)$$

Calculation of internal axial forces acting on the windings of the transformer is presented in Figure 6.9.

You can substitute (6.18) in the formula (6.17) and then we get the expression for the axial force acting on the winding of the power transformer:

$$F_{axial} = -\frac{\mu_0 (I_{max}\omega)^2 \rho_1 \pi D_{12} \delta'}{2(h_{LV}+\Delta)} \tag{6.19}$$

Where

$\delta' = \delta_{12} + \frac{b_1+b_2}{3}$, this is the maximum width of the main leakage channel.

6.2.4. Elliptical Deformation of Winding.

Here we study the case where one or more winding segments have been slightly deformed from the ideal circular form to an elliptic form [by 24] (Figure 6.10).

To the first order of approximation in the small parameter ε/R, we obtain the following radial deformation δ (r, φ) of the elliptic winding compared to the unperturbed circular winding:

$$\delta(r, \varphi) = \delta(\varphi) = r - R = \varepsilon \cos(2\varphi) \tag{6.20}$$

where R - the radius of the undeformed winding, r - the curve describing the elliptical deformation, ε = *R*-*r*(90°).

For small values of the ratio ε/R, the condition of maintaining the length of the coil before and after deformation is satisfied:

$$l = 2\pi R(1 + \frac{\varepsilon^2}{4R^2}) \approx 2\pi R \tag{6.21}$$

Substituting the equation of elliptic deformation into the force formula for radial deformation, we get:

$$F_r = \frac{\mu_0 (I_{макс}\omega)^2 \rho_1 \pi (D_{12} + \varepsilon \cdot \cos(2\varphi))}{2h} \tag{6.22}$$

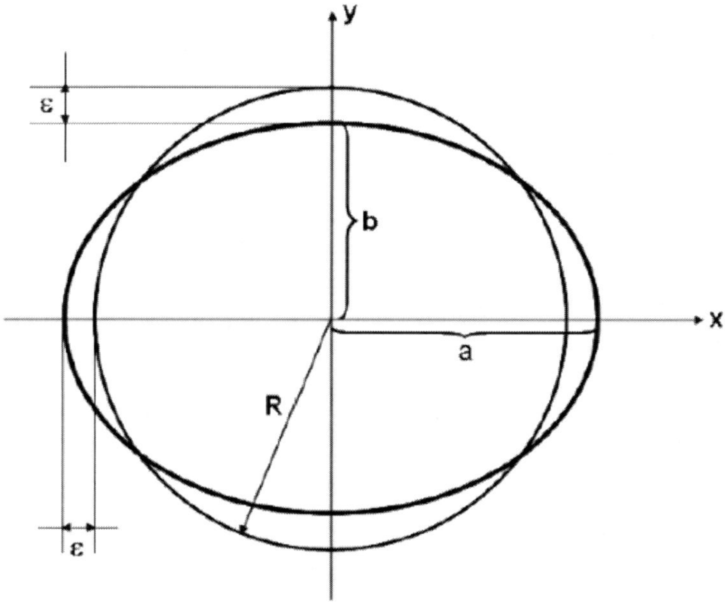

Figure 6.10. Elliptical deformation.

The whole refinement of the formula for radial forces is reduced to the fact that the distance *D12* changes, which is the middle between the primary and secondary windings. Initially, we assume that only one of the windings is deformed, and then the formula correctly reflects the processes occurring in the transformer winding. In the literature [24] about the second winding in the elliptical deformation is not mentioned.

The second assumption with elliptical deformation is that we mean that the ellipse is as the *y* axis perpendicular to the tank wall, but this may not be the case and will have to enter another angle into the equation. But this is a topic for further research.

6.3. Development of the Mathematical Models of the External Diagnostic Action of Pulse on the Schematic of the Substitution of the Windings of the High-Voltage Electrical Apparatuses

According to statistics, for 110-500 kV transformers and a capacity of 63 MVA or more of electrical and inter-system networks of Russia, about 30% of the total number of electrical equipment outages is associated with internal faults [32]. One of the reasons for the internal windings short-circuit of power transformers is the insufficient electrodynamic withstand of the windings at faults, which can lead to breakdown of insulation in the place of residual deformations.

6.3.1. Methods of Diagnostics of Fault Defects of High-Voltage Oil-Filled Transformer-Reactor Equipment

The most effective and widespread methods for diagnosing fault defects of high-voltage oil-filled transformer-reactor equipment are methods, measuring the deviation of the short-circuit impedance value from the reference value or from the values of Zk in other phases, chromatographic analysis of the gases dissolved in the transformer oil (DGA), measuring the intensity of partial discharges (PD) in insulation and other methods.

It should be noted that FRA & LVI-diagnostics is more sensitive to any changes in the mechanical state of transformer & reactor windings - local small changes in the interturn, inter-coil, interwinding, interphase, magnetic circuit and the transformer tank of mutual and own capacitances and inductances, when considering the problems of the electrodynamic stability of the windings of power transformers and reactors at short circuit. The Zk method reacts only to significant changes in the winding inductance, if residual deformations capture a significant number of turns and a breakdown of the electrical strength of the insulation may occur [33-35].

Nevertheless, despite its effectiveness in detecting residual winding deformations, FRA & LVI-diagnostics does not allow quantifying the residual deformations that have occurred, requires tripping and removal of busbars of the transformer. The results of FRA & LVI measurements may depend on the measurement scheme (the mutual arrangement of the measuring cables, the influence of the busbar near the object of measurements, external interferences in the conditions of the operating high-voltage substation and other factors) [36].

According to the oscillograms of FRA & LVI, only a qualitative assessment of the deformations is possible, and it is not always possible to fully integrate changes in the FRA & LVI signals with complete certainty. Therefore, the task was set to proceed to a quantitative evaluation of the results of FRA & LVI. It can be achieved by developing mathematical models of the external diagnostic impact of an aperiodic pulse on the winding electrical scheme of substitution and estimating the values of the above capacitances and inductances, and determining their changes in the winding during its damage [36].

The questions of operating modes, features of the magnetic field and the use of calculations of increased accuracy for evaluating the state of windings of power transformers are considered in [36-38].

6.3.2. Development of Mathematical Models of External Diagnostic Impact

Development of mathematical models of external diagnostic impact on the circuit for the replacement of windings of high-voltage electrical apparatus (EA) for the purpose of quantitative evaluation of the results of diagnostic measurements, for development of an electrical scheme of substitution for the diagnosed reactor winding (transformer).

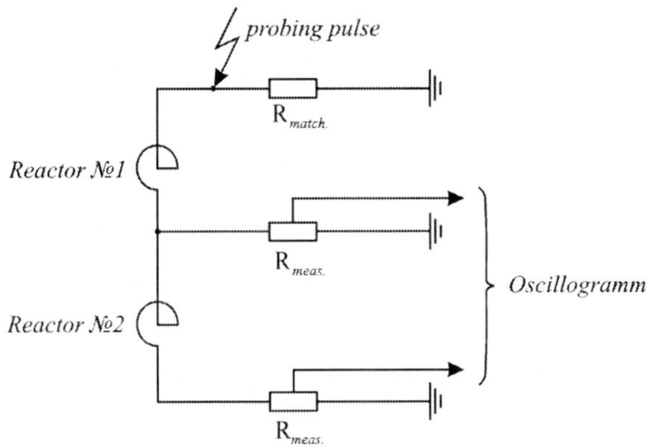

a)

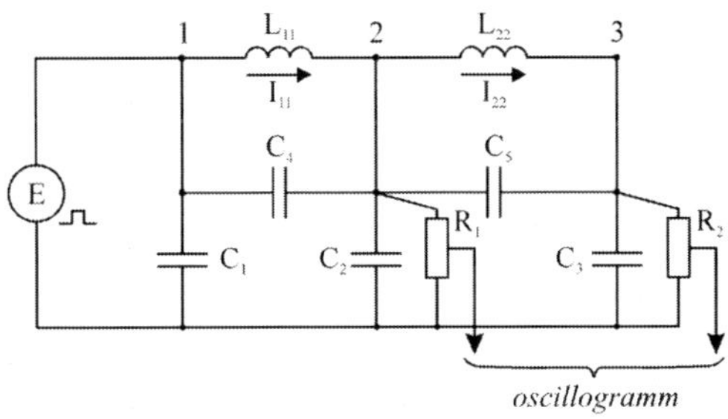

b)

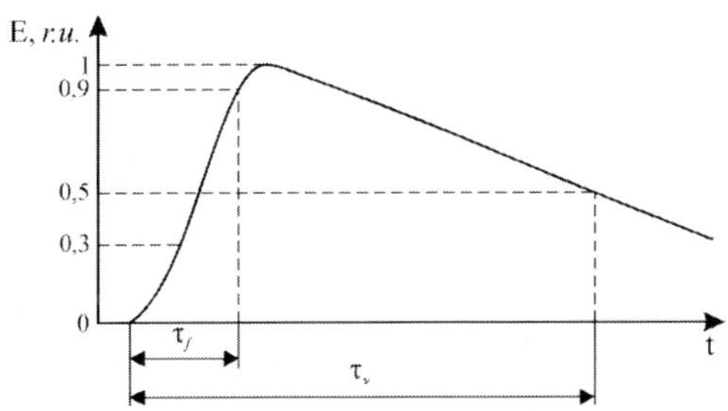

c)

Figure 6.11. Scheme of measurement under external diagnostic impact of an aperiodic impulse on the electrical winding scheme of substitution of two reactors: a) scheme for FRA & LVI measuring, b) scheme of substitution for reactors, c) the form of an aperiodic probing pulse.

In the framework of this task, it is also necessary to develop the appropriate mathematical and software to determine the real parameters of the EA winding by FRA & LVI signals for estimating the change in its parameters (inductance, capacitance), i.e., geometric dimensions of the winding after the passage of short-circuit currents.

In the [40, 41], the case of free oscillations in a winding without allowance for mutual induction between turns is considered when an aperiodic pulse is applied to it. Let us consider this with reference to the diagnostic impact of an aperiodic pulse on the windings of two reactors. The measurement scheme is shown in Figure 6.11a, the electrical scheme of substitution of two reactors in Figure 6.11b.

In the reactor electrical scheme of substitution, we have:

L_{11}, L_{22} - own reactors inductance,

C_1, C_2, C_3 - capacities of reactors on the ground,

C_4, C_5 - longitudinal capacities between reactors.

Let us take the following assumptions in order to facilitate calculations:

1) measuring resistance $R_1 = R_2 = \infty$;
2) E - is an aperiodic impulse;
3) we neglect active resistances;
4) we neglect mutual inductances.

Let us write in a general form the system of equations for all branches:

$$\|I_B\| = \|Y_B\| \bullet \left(\|\Delta U\| + \|E\|\right) \tag{6.23}$$

Then for 3 points 1, 2, 3 in Figure 6.11b we have:

$$\begin{Vmatrix} I_1 \\ I_2 \\ I_3 \end{Vmatrix} = \begin{Vmatrix} Y_1 0 0 \\ 0 Y_2 0 \\ 0 0 Y_3 \end{Vmatrix} \bullet \left(\begin{Vmatrix} \Delta U_1 \\ \Delta U_2 \\ \Delta U_3 \end{Vmatrix} + \begin{Vmatrix} E_1' \\ E_2' \\ E_3' \end{Vmatrix} \right), \tag{6.24}$$

where $E_1^{'}$, $E_2^{'}$, $E_3^{'}$ - obtained for each point number 1, 2, 3 after the transformation of the "star" into the "triangle" of the source EMF;

Y_1 , Y_2 , Y_3 - own conductivities of the branches, the zeros in the conductivity matrix are in place of mutual conductivities, which we neglect;

ΔU_1, ΔU_2, ΔU_3 - respectively, the voltage drops on the branches..

We apply the operator method. We write the solution for the potential at the nodes 1, 2, 3. In the operator form, this is a system of linear algebraic equations with respect to $U_i(p)$:

$$U_i(p) = \frac{\Delta i(p)}{\Delta(p)} \bullet E(p), \tag{6.25}$$

where 1, 2, 3 - the schemes in Figure 6.11b;

$\Delta(p)$ - the determinant of the conductivity matrix from (6.24);

$\Delta i(p)$ - determinant obtained from $\Delta(p)$ when the i-th column is replaced by the column of the source matrix.

When two reactors are subjected to an aperiodic probing pulse:

$$E(p) = \frac{1}{p}, \tag{6.26}$$

then from (6.25) we have:

$$U_{i0}(p) = \frac{\Delta i(p)}{\Delta(p)} \bullet \frac{1}{p}. \tag{6.27}$$

We perform the inverse Laplace transform of (6.27):

$$U_{i0}(t) = \frac{\Delta i(0)}{\Delta(0)} + \sum_{k=1}^{4} \frac{\Delta i(p_k)}{p_k \Delta^{'}(p_k)} \bullet e^{p_k t}, \tag{6.28}$$

where $\Delta^{'}(p_k) = \frac{d\Delta(p)}{dp}|p = p_k$; (6.29)

when p_k - the roots of the determinant $\Delta(p)$, these are own numeric of the reactors.

Since the scheme without active resistances is considered, all the roots p_k of equation $\Delta(p) = 0$ will be imaginary.

The module p_k will be the natural frequency ω_k of the winding, i.e.

$$p_k = \pm j\omega_k .$$

Equation (6.28) will have the form:

$$U_{i0}(t) = \frac{\Delta i(0)}{\Delta(0)} + 2\,\mathrm{Re}\sum_{k=1}^{4} \frac{\Delta i(j\omega_k)}{j\omega_k \Delta^{'}(j\omega_k)} \bullet e^{j\omega_k t} \quad (6.30)$$

In the expression $\Delta(p)$ and $\Delta i(p)$

Argument ***p*** enters only into an even power, but into a polynomial $\Delta^{'}(p)$, which does not have a free term, only in an odd degree.

Equation (6.30) will have the form:

$$U_{i0}(t) = \frac{\Delta i(0)}{\Delta(0)} + 2\sum_{k=1}^{2} \frac{\Delta^{*} i(\omega_k)}{\omega_k \Delta^{*'}(\omega_k)} \bullet \cos\omega_k t \quad (6.31)$$

or

$$U_{i0}(t) = \frac{\Delta i(0)}{\Delta(0)} + 2\left[\frac{\Delta^{*} i(\omega_1)}{\omega_1 \Delta^{*'}(\omega_1)} \bullet \cos\omega_1 t + \frac{\Delta^{*} i(\omega_2)}{\omega_2 \Delta^{*'}(\omega_2)} \bullet \cos\omega_2 t\right] \quad (6.32)$$

where - $\overset{*}{\Delta} i(\omega_k)$ and $\overset{*'}{\Delta}(\omega_k)$ - determinants (polynomials) analogous to $\Delta i(\omega_k)$ and $\Delta'(\omega_k)$, but in the calculation it is taken into account that

$$p_k = \pm j\omega_k$$

In expression (6.32), the second term in square brackets depends on time and represents free oscillations in the winding of the reactor, and the first term is independent of time and represents a quasi-stationary voltage $U_{i\_к.с.}$ at a given point in measurement scheme, i.e.

$$U_{i\_к.с.} = \frac{\Delta i(0)}{\Delta(0)} \tag{6.33}$$

Thus, we have:

$$U_{i0}(t) = U_{i\_к.с.} + 2\left[\frac{\overset{*}{\Delta} i(\omega_1)}{\omega_1 \overset{*'}{\Delta}(\omega_1)} \bullet \cos\omega_1 t + \frac{\overset{*}{\Delta} i(\omega_2)}{\omega_2 \overset{*'}{\Delta}(\omega_2)} \bullet \cos\omega_2 t\right] \tag{6.34}$$

In our case it is convenient to use the Duhamel integral in the form:

$$U_i(t) = E(t)U_{i0}(t) + \int_0^t E(t-\tau)\frac{dU_{i0}(\tau)}{d\tau}d\tau \tag{6.35}$$

The aperiodic probing pulse can be represented in the form (see Figure 6.11c):

$$E(t) = E_0(e^{-\alpha t} - e^{-\beta t}), \tag{6.36}$$

when E_0 - multiplier, which brings the amplitude of probing pulse to unity, since it is convenient to carry out the calculation in relative units;

α, β - parameters determined by the waveform of the pulse. Substituting (6.36) and (6.34) into (6.35) we obtain:

$$U_i(t) = U_{i\_к.с.} E_0 \left[e^{-\alpha t} - e^{-\beta t}\right] + 2E_0 \left[\frac{\Delta^* i(\omega_1)}{\omega_1 \Delta^{*'}(\omega_1)} \bullet \frac{\left(\beta^2 - \alpha^2\right)\omega_1^2}{\left(\alpha^2 + \omega_1^2\right)\left(\beta^2 + \omega_1^2\right)} \bullet \right.$$
$$\bullet \cos\omega_1 t + \frac{\omega_1 \left(\beta - \alpha\right)\left(\omega_1^2 - \alpha\beta\right)}{\left(\alpha^2 + \omega_1^2\right)\left(\beta^2 + \omega_1^2\right)} \bullet \sin\omega_1 t + \frac{\alpha^2}{\alpha^2 + \omega_1^2} e^{-\alpha t} -$$
$$- \frac{\beta^2}{\beta^2 + \omega_1^2} e^{-\beta t} + \frac{\Delta^* i(\omega_2)}{\omega_2 \Delta^{*'}(\omega_2)} \bullet \frac{\left(\beta^2 - \alpha^2\right)\omega_2^2}{\left(\alpha^2 + \omega_2^2\right)\left(\beta^2 + \omega_2^2\right)} \bullet \cos\omega_2 t +$$
$$\left. + \frac{\omega_2 \left(\beta - \alpha\right)\left(\omega_2 - \alpha\beta\right)}{\left(\alpha^2 + \omega_2^2\right)\left(\beta^2 + \omega_2^2\right)} \bullet \sin\omega_2 t + \frac{\alpha^2}{\alpha^2 + \omega_2^2} e^{-\alpha t} - \frac{\beta^2}{\beta^2 + \omega_2^2} e^{-\beta t} \right] \quad (6.37)$$

6.3.3. Expression for Describing the Change in the Mechanical State of the Winding

Expression (6.37) allows to present FRA or LVI oscillogram after the winding changes of mechanical state due to the flow of short-circuit currents, according to the absolute single-phase measurement scheme. To determine FRA or LVI oscillogram according to a differential scheme, i.e., scheme when subtracting signals of two phases, for example “A-B”, to increase the sensitivity of the measurement scheme, you can use the following expression:

between points i and $i+1$,

$$\Delta U_i(t) = U_i(t) - U_{i+1}(t) \quad (6.38)$$

We rewrite (6.37) in the following form:

$$U_i(t) = U_{i\_\kappa.c.} E_0 \left(e^{-\alpha t} - e^{-\beta t}\right) + 2E_0 \left(A_1 \cos\omega_1 t + B_1 \sin\omega_1 t + C_1 e^{-\alpha t} + D_1 e^{-\beta t} + \right.$$

$$\left. + A_2 \cos\omega_2 t + B_2 \sin\omega_2 t + C_2 e^{-\alpha t} + D_2 e^{-\beta t}\right), \tag{6.39}$$

where the coefficients are:

$$\left.\begin{aligned}
A_1 &= \frac{\Delta^* i(\omega_1)}{\omega_1 \Delta^{*'}(\omega_1)} \bullet \frac{\left(\beta^2 - \alpha^2\right)\omega_1^2}{\left(\alpha^2 + \omega_1^2\right)\left(\beta^2 + \omega_1^2\right)}; \\
B_1 &= \frac{\omega_1 \left(\beta - \alpha\right)\left(\omega_1^2 - \alpha\beta\right)}{\left(\alpha^2 + \omega_1^2\right)\left(\beta^2 + \omega_1^2\right)}; \\
C_1 &= \frac{\alpha^2}{\alpha^2 + \omega_1^2}; C_2 = \frac{\alpha^2}{\alpha^2 + \omega_2^2}; \\
D_1 &= \frac{\beta^2}{\beta^2 + \omega_1^2}; D_2 = \frac{\beta^2}{\beta^2 + \omega_2^2}
\end{aligned}\right\} \tag{6.40}$$

Similarly for A_2 and B_2 .

6.3.4. Selection of Optimal Values of the Parameters of the Substitution Circuit

For electrical winding scheme of substitution in Figure 6.11b to correspond well to FRA or LVI oscillogram, i.e., the response of the winding to a rectangular probing pulse, it is necessary to choose the optimum values of the winding scheme parameters on which the shape of FRA or LVI oscillogram depends, i.e., solve the optimization task.

Optimization parameters-variables L_{11}, L_{22}, C_1, C_2, C_3, C_4, C_5.

Target function - voltage curve $U_i(t)$ by (6.39) or according to (6.38),

Limitations - instantaneous voltage values $U_i(t)$ should not differ by more than $\delta U_i(t) = \pm 40\%$ from instantaneous values $U_i^*(t)$ of FRA or LVI oscillograms [43-45].

The optimization method is a search through the values of the above parameters of optimization.

Criterion of optimization - criteria for minimizing of root mean square deviation $\sum$ of amplitude values of the voltage curve $U_i(t)$ from $U_i^*(t)$:

$$\Sigma = \sqrt{\sum_i \left[U_i(t_j) - U_i^*(t_j)\right]^2}, \tag{6.41}$$

where the index j corresponds to the amplitude values of the voltage curve $U_i(t_j)$ and $U_i^*(t_j)$ [40, 41].

Calculation of optimal parameters of variables L_{11}, L_{22}, C_1, C_2, C_3, C_4, C_5 is performed when the system of partial differential equations of the square root mean square deviation $\sum$ is jointly solved by optimization parameters:

$$\left.\begin{aligned} &\frac{\delta\Sigma^2}{\delta L_{11}} = 0; \frac{\delta\Sigma^2}{\delta L_{22}} = 0; \frac{\delta\Sigma^2}{\delta C_1} = 0; \\ &\frac{\delta\Sigma^2}{\delta C_2} = 0; \frac{\delta\Sigma^2}{\delta C_3} = 0; \frac{\delta\Sigma^2}{\delta C_4} = 0; \frac{\delta\Sigma^2}{\delta C_5} = 0 \end{aligned}\right\} \tag{6.42}$$

CONCLUSION

1. The electromagnetic force acting on a winding is directly proportional to the winding current and the flux density.
2. The electromagnetic force acting on an element of a current in a magnetic field can be determined according to the law of Bio-Savart.
3. The radial forces acting on the concentric windings of a two-winding transformer tend to increase the distance between the windings.
4. Equation for deformations of LV winding of 250MVA/220 kV transformer after short-circuit testing with the loss of the radial stability (radial buckling) of winding is given.
5. The resulting expression for the radial force more accurately reflects the processes occurring in the windings of power transformers in case of short circuit and the occurrence of radial hoop buckling, due to the use of trigonometric functions in describing the shape and volume of deformation.
6. Calculation of internal axial forces acting on the windings of the transformer was done.
7. According to statistics, about 30% of the total number of outages of electrical equipment is associated with internal faults.
8. The task of the transition to a quantitative assessment of the results of FRA or LVI was set and solved by developing mathematical models of the external diagnostic impact of an aperiodic pulse on electrical winding scheme of substitution.
9. Mathematical software has been developed to determine real parameters of transformer & reactor winding by the FRA or LVI oscillograms for estimating the change in its parameters (inductance, capacitance).
10. The impact of an aperiodic pulse on the windings of two reactors is considered.
11. The problem of optimization parameters for the circuit for replacing the windings of two reactors is solved.Further development of the

circuit for replacing the diagnosed reactor and the corresponding mathematical and software software can allow to determine the real parameters of the reactor from the LVI pulse current curve, which will make it possible to judge the changes in the reactor winding parameters by changes in the LVI curve.

ACKNOWLEDGMENTS

Dr. Alexei Tadzhibaev, Saint Petersburg Institute of Electrical Engineering additional education; Dr. Oleg Shlegel (his memory), scientific supervisor of my master's (candidate) thesis; Mr. Nikolaj M. Alexandrov, post-graduate student, Russia, JSC "Dinamika" greatly acknowledged for supporting this study. Cooperation of Universities and Innovation Development, Doctoral School project "Complex diagnostic modeling of technical parameters of power transformer-reactor electrical equipment condition", 2009, has made publishing of this article possible.

REFERENCES

[1] Mohd Fairouz Bin Mohd, Y. (2015). *Frequency Response Analysis for Transformer Winding Condition Monitoring.* A thesis submitted for the degree of Doctor of Philosophy. The University of Queensland, Australia.

[2] Janssen, A.L.J., te Paske, L.H., Smeets, R.P.P., Leufkens, P.P., Fogelberg, T. (2009). Years test experience with short-circuit withstand capability of large power transformers. *CIGRÉ 2009, 6th Southern Africa Regional Conference,* Paper P501.

[3] Lapworth, J.A. (2001). Recent Developments Relating to the Detection of Winding Movement in Transformers by Frequency Response Analysis (FRA). *Sixty-Seventh Annual International Conference of Doble Clients.*

[4] Bertagnolli, G. (1998). Short-Circuit Duty of Power Transformers. *ABB Transformatori*, Legnano (Milano) – Italy. 197.

[5] IEEE Std C57.12.90-1993, Part II: *IEEE Guide for Short-Circuit Testing of Distribution and Power Transformers.*

[6] Janssen, A.J.L., te Paske, L.H. (2000): Short-Circuit testing experience with large power transformers, *CIGRE SC12 Session 2000, report 12–105.*

[7] Janssen, A.L.J., te Paske, L.H., Smeets, R.P.P. (2002). Short-circuit Tests on Large Transformers: Destructive or not?, *TrafoTech* 2002, Mumbai, India, 126-33.

[8] Janssen, A.L.J., te Paske, L.H., van der Linden, W.A., Smeets, R.P. P. (2001). *Magnetic Saturation at Short-Circuit Tests on Power Transformers,* IEEE PES Summer Meeting, paper 72.03.

[9] Bertagnolli, G. (1996). *Short-Circuit Duty of Power Transformers,* Book printed on behalf of ABB Trasformatori Legano (Milano), ISBN Italy.

[10] IEC Committee Draft 14/465/CD Annex B (2003). *Theoretical Evaluation of the Ability to Withstand the Dynamic Effects of Short Circuit.*

[11] CIGRE WG12.05 (1985). An International Survey on Failures in Large Power Transformers in Service, *Electra* 88, 21–37.

[12] Bertagnolli, G. (1999). Results of short-circuit performance of transformers, *CIGRE Transformer Colloquium,* Budapest, Short-circuit performance; Tests and Failures, report 2, CIGRE TC 12.

[13] Macor, Robert, G., Girardot, D., Riboud, J.C., Ngegueu, T., Arthaud, J.P., Chemin, E. (2000). The Short-Circuit Resistance of Transformers: The Feedback in France Based on Tests, Service and Calculation Approaches. *CIGRE Conference,* paper 12–102.

[14] Khrennikov, A.Yu. (2004). Short-circuit performance of power transformers. Test experience, including fault diagnostic, *Reports of Advanced Research Workshop on Modern Transformers* (ARWtr 2004), Vigo – Spain.

[15] Khrennikov, A.Yu. (2006). Diagnostics of electrical equipment's faults, defects and weaknesses, *Reports of Conference on Condition Monitoring and Diagnosis* (CMD 2006), Korea.

[16] Khrennikov, A.Yu., Mazhurin R.V., Radin P.S., Skrydlov E.I., Khrennikova T.A. Estimation of the technical condition state of substation electrical equipment with the use of software and information tools. *Journal of Basic and Applied Research International,* 2015-Volume 7 (Issue 3), – P. 201-210. .http://www.ikpress.org/issue.php?iid=520&id=42.

[17] Khrennikov, A.Yu. (2012). Monitoring information-measuring system for detecting power transformer faults, FRA and LVI-testing diagnostics experience, *Reports of Asia-Pacific Power and Energy Engineering Conference (APPEEC 2012)*, Shanghai, China.

[18] Khrennikov A.Yu. New "intellectual networks" (Smart Grid) for detecting electrical equipment faults, defects and weaknesses. *Smart Grid and Renewable Energy,* February, 2012, Volume 3, Issue 1.

[19] Khrennikov, A.Yu. (2013). Smart Grid technologies for Detecting Electrical Equipment Faults, Defects and Weaknesses. *Workshop on Mathematical Modelling of Wave Phenomena with applications in the power industry*, Linnaeus University, Växjö. http://lnu.se/mmwp.

[20] Khrennikov, A.Yu., Shakaryan, Y.G., Dementyev, Y.A. (2013). Shortcurrent Testing Laboratories. Short-Circuit Performance of Power Transformers, Transformer Testing Experience. *International Journal of Automation and Control Engineering (IJACE),* Volume 2, Issue 3. (pp. 120-127). http://www.seipub.org/ijace/AllIssues.aspx.

[21] Khrennikov, A. Yu. (2014). Transformer Testing Experience by LVI/FRA Methods. Short-Circuit Performance of Power Transformers. *Current Advance in Energy Research (CAIER).* Volume 1, Issue 1 Jun. (pp. 8-14). www.vkingpub.com/VkUpload/201406131101418881.pdf.

[22] Khrennikov, A.Yu., Mazhurin, R.V. (2014). The SF6 high-voltage electrical equipment: analysis of accident rate and service experience. *Current Advance in Energy Research (CAIER)* Volume 1, Number 2. (pp. 44-49). http://www.vkingpub.com/Journal/CAIER/245.html.

[23] Khrennikov, A.Yu., Mazhurin, R.V. (2014). High-voltage electrical equipment with the SF6 (sulfur hexafluoride) gas insulation: analysis of accident rate and service experience. *International Journal of Engineering Research & Management (IJERM),* Volume 1, Issue 5. (pp. 31-36). http://www.ijerm.com/ar_download.php?archive=Vol.1+Issue+5&&archive_id=9.

[24] Dalarsson, M. (2013). Using Multiple Modes to Reconstruct Conductor Locations in a Cylindrical Model of a Power Transformer Winding. *Workshop on Mathematical Modelling of Wave Phenomena with applications in the power industry*, Linnaeus University, Växjö. http://lnu.se/mmwp.

[25] Maldaque, X. (2001). *Theory and Practice of Infrared Technology for Nondestructive Testing,* John Wiley & Sons, Inc., New York.

[26] Vavilov, V., Burleigh, D. (2001). *Nondestructive testing handbook, ASNT,* Vol.3, 54-75.

[27] Hanson, C. (2011). *Infrared Products,* Garland, L-3 Communications Electro-Optical Systems.

[28] Chou, Y., Yao, L. (2009). Automatic diagnostic system of electrical equipment using infrared thermography, *Proceedings of International Conference of Soft Computing and Pattern Recognition*, 155-160.

[29] Lizak, F., Kolcun, M. (2008). Improving reliability and decreasing losses of electrical system with infrared thermography", *Acta Electrotechnica et Informatica,* Vol. 8, No. 1, 60-63.

[30] Andonova, A. (2012). Thermographic evaluation of electro-mechanical relays' quality in railway automation, *International Journal of Electrical and Computer Engineering (IJECE),* vol.2, No1, 1-6.

[31] Patent of USA, US 6446027 B1, O'Keeffe, Thomas Gary (2002). *Intelligent Analysis System and Method for Fluid-Filled Electrical Equipment.*

[32] Lvov, M.Yu., Lvov, Yu.N., Dement'ev, Yu.A., Antipov, K.M., Surba, A.S., Sheiko, P.A., Neklepaev, B.N, Shifrin L.N., Kassikhin S.D., Slavinsky A.Z., Sipilkin K.G. About reliability of power transformers

and autotransformers of electric networks. *Electrical stations.*- № 11.- 2005 (in Russian).

[33] Drobyshevsky, A.A., Levitskaya, E.I. Quantitative estimation of the results of pulsed defectography of windings of power transformers. *Electrical Engineering,* 1990, № 5(in Russian).

[34] Khrennikov, A. Yu., Kikov, O.M. Diagnostics of power transformers in Samaraenergo by the method of low-voltage pulses. *Electrical stations,* № 11, 2003 (in Russian).

[35] Khrennikov, A. Yu., Schlegel, O.A. Monitoring the change in the inductive reactance of the transformer to determine the damage in the windings. *Energetik,* № 2, 2004 (in Russian).

[36] Khrennikov, A.Yu. Methods of low-voltage pulses and frequency analysis for monitoring the mechanical state of windings of power transformers. *Electro,* № 2, 2007 (in Russian).

[37] Alexandrov, G.N. Features of the transformer magnetic field under load. *Electricity,* 2003, № 5 (in Russian).

[38] Alexandrov, G.N. Transformer operation modes. *Manual, St. Petersburg, Center for the Training of Power Engineering*, 2005, 141 p. (in Russian).

[39] Alpatov, M.E., Bazutkin, V.V. Use of calculations of increased accuracy for an estimation of a condition of windings of power transformers. *ELECTRO,* № 5, 2005 (in Russian).

[40] Geller, B., Verevka, A. Pulsed processes in electric machines. Moscow: *Energy,* 1973. - 439p. (in Russian).

[41] Lohanin, A.K. Calculation of overvoltages in the coil windings of transformers. *Electricity*, 1967, № 4 (in Russian).

[42] Khrennikov, A.Yu. The main causes of damage to windings of power transformers in case of short circuits. *Electricity,* 2006, № 7 (in Russian).

[43] Khrennikov, A.Yu. The main causes of winding damage of power transformers with a voltage of 110-500 kV during operation. *Industrial Energy,* 2006, № 12.-p.12-14 (in Russian).

[44] Khrennikov, A.Yu. About reliability and methods of diagnostics of high-voltage electrical equipment of substations. *New in the Russian Electrical Power Industry,* № 7, 2006, p.35-45 (in Russian).

[45] Khrennikov, A.Yu., Tadzhibaev, A.I. Methods for assessing the state of power oil-filled transformers based on the control of winding geometry. *Monograph*, PEIPK, St. Petersburg, 2005, 50 p. (in Russian).

Chapter 7

PERSPECTIVE FOR USING PHOTOTHYRISTORS, IGCT AND IGBT FOR ELECTRODYNAMIC TESTING OF POWER TRANSFORMERS

A. A. Kuvshinov*[*], *PhD and A. Yu. Khrennikov, PhD

ABSTRACT

A comparative characteristic of the switching capabilities of photothyristors, lockable thyristors IGCT and high-power transistors IGBT is given, taking into account the specific operating conditions in the composition of the high-voltage high-current semiconductor key (HHCSK) during electrodynamic tests. The necessity of using a damping circuit as a part of a HHCSK at any type of used power semiconductor device for compulsory formation of the required trajectory of the recovering voltage is grounded. A safety criterion for electrodynamic tests is proposed, which consists in the forced cessation of the short circuit experience immediately after the formation of an emergency state signal for the windings of the tested power transformer. It is shown that the

[*] Corresponding Author's E-mail: alekseikuvshinov@yandex.ru.

practical implementation of the proposed criterion is possible only with the help of HHCSK on fully controlled power semiconductor devices. The protection circuit for overvoltage of the tested power transformer and HHCSK in case of emergency interruption of the short circuit current is considered. Presents examples of the implementation of the power circuit HHCSK on lockable thyristors with an integrated control device (IGCT) and insulated gate bipolar transistors (IGBT).

Keywords: power transformer, electrodynamic tests, power semiconductor device, high-voltage high-current semiconductor key

7.1. INTRODUCTION

The organizational and technical problems of conducting electrodynamic tests of power transformers have already been considered in a number of works [1-7]. It is emphasized that the main difficulties are related to the requirements for accurate synchronization of the beginning and end of short-circuit tests, which result from the need for rationing and repeatability of levels of influence on power transformers [8-15]. This requires the use of specialized high-voltage switches with controlled commutation and a traditional contact system [6] or high-voltage heavy current semiconductor keys (HHCSK) based on power semiconductor devices (PSD). The latter have a much greater switching resource and synchronization accuracy.

The practice of carrying out electrodynamic tests using high-voltage thyristor valves (HTV) of a wide range of power transformers rated up to 666 MVA and rated voltage 110¸ 750 kV is known [16]. Such tests were carried out at Power Short-circuit Laboratory, Togliatti. Features of operating conditions and circuit technology HTV, due to the specifics of conducting electrodynamic tests, are considered in [3, 6, 14, 16].

At present, on the PSD market, in addition to traditional low-frequency thyristors with electric current control, powerful low-frequency thyristors with direct light control (LTT), lockable thyristors with a ring lead of the control electrode and an integrated control device (IGCT), as well as powerful bipolar transistors with isolated shutter (IGBT). By the amount of switched power, these devices are quite comparable with traditional low-frequency thyristors, and even exceed the dynamic characteristics and functionality.

Therefore, it is quite logical to formulate the problem of a comparative analysis of the HHCSK power circuits realized using various types of PSD. This is necessary for a technically reasoned choice of the most acceptable version of the HHCSK implementation [3, 6, 14, 16-20].

7.2. Functional and Commutation Special Features of High-Voltage Heavy Current Semiconductor Key (HHCSK) for the Electrodynamic Tests of the Transformer

7.2.1. Requirements for the Functionality of Commutation Equipment

Electrodynamic tests are an important and necessary stage in the development of new transformer and reactor equipment, which verifies the adequacy of implemented design and technological solutions. To make full use of the potential of such tests, switching equipment is required that can provide [14, 16, 18]:

- several cycles of switching on/off of the tested power transformers with artificially shorted terminals to the test bus feed rails with the required short-circuit capacity;
- synchronization of the beginning of the short-circuit tests with the voltage on the busbars of the test stand with an accuracy of one microsecond to ensure the full repeatability of the experiments;
- regulation of the start-up phase in the range ($0 \div \pi / 2$) with respect to the moment of voltage transition on the test-bench buses through

the zero value for rationing of the aperiodic component of the short-circuit current;

- synchronization of the end of the experiments with the moment of the short-circuit current transition through a zero value to minimize overvoltages;
- preliminary magnetization of the magnetic circuit of the tested transformer with a unipolar current from the voltage on the busbars of the test bed;
- regulation of the duration of the short-circuit tests from one period of the industrial frequency to the maximum permissible value (3sec. in accordance with National Standard GOST 11677-85).

It is practically impossible to ensure the fulfillment of the above requirements by means of contact switching devices (high-voltage switches), since the switching-on and shut-off times depending on the specific type can be tens of milliseconds and are statistical characteristics. In addition, high-voltage switches have a limited switching resource.

Thus, there is a real need to create a high-voltage heavy current semiconductor key (HHCSK) for conducting short-circuit tests with a normalized level of effects during electrodynamic tests of power transformers. The modern elemental base of power electronics provides a wide enough choice of various types of power semiconductor devices (PSD), which have the potential to be used in the HHCSK:

- high-power low-frequency thyristors with electric current control (ETT), the features of their operation under the conditions of short-circuit tests of power transformers were considered in [4];
- powerful low-frequency thyristors with direct light control (phototyristors - LTT), the use of which greatly simplifies the circuit and design of a high-voltage thyristor valve;
- lockable thyristors (GTO), which are switched on and off by applying positive and negative current pulses to the control electrode;

- lockable thyristors with a ring lead of the control electrode, called thyristors with rigid control (GCT) and more advanced modifications - lockable thyristors with an integrated control device-driver (IGCT);
- bipolar transistors with isolated gate (IGBT).

In terms of functionality, each of these PSDs is able to provide a set of requirements formulated above. However, when choosing the type of PSD, it is necessary to take into account the specific features of the operating mode of the HHCSK during electrodynamic tests. The main feature is that the emergency mode of short-circuit of the tested transformer is essentially nominal for the HHCSK. The latter must have the appropriate switching capacity and switching resource in order to repeatedly pass the short-circuit currents of a wide range of power transformers without exceeding the maximum permissible values of the parameters of the PSD. For this, the short-duration of the short-circuit tests must be taken into account [14, 16, 18-25].

7.2.2. Requirements for the Commutation Special Features of HHCSK

The overload capability of the HHCSK should be sufficient to pass the short-circuit current on the high-voltage side of the power transformer under test when the terminals on the low-voltage side are short-circuited. The multiplicity of the steady-state short-circuit current in the two-winding regimes of a three-phase transformer is given by

$$I_{SC}^{*} = \frac{I_{SC}}{I_{NOM}} = \frac{1}{(u_{SC\%}/100 + S_{NOM}/S_C)}, \tag{7.1}$$

where $u_{SC\%}$is the short-circuit voltage of the transformer under test, expressed as a percentage; I_{NOM}, S_{NOM} – rated current and rated power of the tested power transformer; S_C – short-circuit power of the mains on the busbars of the test stand (taking into account the intermediate transformers for matching the voltages on the busbars of the test stand with the rated voltage of the tested transformer).

For power transformers with rated power $S_{NOM} = (40 \div 1250) MVA$, rated voltage of network windings $(110 \div 500)$ *kV* and short-circuit voltage $u_{SC\%} = (10 \div 14)\%$ at $S_C \approx 0$ (infinite power bus), the multiplicity of the steady-state short-circuit current is $I^*_{SC} = (7{,}14 \div 10)$. In named units, the effective values of the steady-state short-circuit current are $\approx (1 \div 18)$ *kA*.

In general, electrodynamic tests may provide for the effect on the test transformer of a steady-state short-circuit current with the application of an aperiodic component with a normalized value of the shock coefficients. According to GOST 11677-85, the maximum short-circuit current (the amplitude of the first half-wave of the short-circuit current)

$$I_{sh_m} = I_{SC} \cdot \left(\sqrt{2} \cdot k_{sh}\right)$$

is determined by the magnitude of the shock coefficient k_{sh}, which, depending on the ratio $u_{SC(r)}/u_{SC(a)}$ of the reactive $u_{SC(r)}$ and active $u_{SC(a)}$ components of the short-circuit voltage of the transformer under test, can be in the range $k_{sh} = (1{,}07 \div 1{,}8)$. For high-power transformers take the highest possible value $k_{sh} = 1{,}8$.

For comparative estimations of HHCSK realized on the basis of various types of PSD, it is advisable to use the notion of switching power of the HHCSK, which means the product of the effective value of the blocking voltage for the effective value of the transmitted current. It is necessary to distinguish the nominal switching power, the magnitude of which is determined by the steady-state short-circuit current, and the shock switching power, the magnitude of which is determined by the shock current of the short circuit.

Expression (7.1) makes it possible to estimate the value of the rated switching power when the steady-state short-circuit current is passed and the voltage on the busbars of the test stand is equal to the rated voltage of the network winding of the tested transformer

$$S^*_{SC_{NOM}} = \frac{S_{SC_{NOM}}}{S_{NOM}} = \frac{I_{SC} \cdot U_{NOM}}{\sqrt{3} \cdot S_{NOM}} = \frac{1}{3 \cdot (u_{SC\%}/100 + S_{NOM}/S_C)}. \quad (7.2)$$

In the most unfavorable case of $S_{NOM}/S_C \approx 0$ (the test bench of infinite capacity) and at typical values of short-circuit voltage $u_{SC\%} = (10 \div 14)\%$, the rated switching power of the HHCSK reaches $S^*_{SC_{NOM}} \cong (2{,}38 \div 3{,}33)$, that is, more than $(2 \div 3)$ times the nominal power S_{NOM} of the power transformer under test. Shock switching power due to the aperiodic component of the short-circuit current increases to $S^*_{SC_{sh}} = k_{sh} \cdot S^*_{SC} \cong (4 \div 6)$. Thus, in order to provide electrodynamic testing of power transformers with rated power up to $S_{NOM} = 1250$ MVA, the shock switching power of the HHCSK should be $S_{SC_{sh}} \approx 7500$ MVA.

The most powerful serial PSD tablets with a diameter of a silicon wafer $(100 \div 125)$ mm are able to switch power only up to $(10 \div 20)$ MVA. This is many times less than the required switching power of the HHCSK. This circumstance must be taken into account when choosing the type of PSD, the power scheme of the HHCSK, and the way it is used in short-circuit tests [14, 16, 18].

7.2.3. Inclusion Schemes of HHCSK

Figure 7.1 presents two possible options for the inclusion of HHCSK in the scheme of conducting the electrodynamic tests of the power transformer T.

In the first case (Figure 7.1a) for the experimental short circuit HHCSK should carry out a full switching cycle consisting of the following stages:

- blocking voltage U_C on the busbars of the test stand after switching on the switch Q1;
- the inclusion of a voltage U_C in the specified phase to ensure the normalized value of the aperiodic component of the short-circuit current;

- passing the fault current of the short circuit during the specified duration of the short circuit test;
- switching off at the instant of short-circuit current transition through zero value to minimize the amplitude of the recovering voltage;
- blocking the recovering voltage before switching off the Q1 switch.

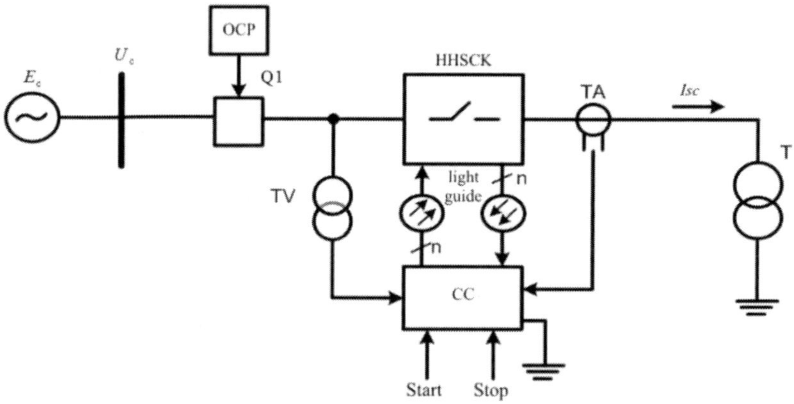

(a)

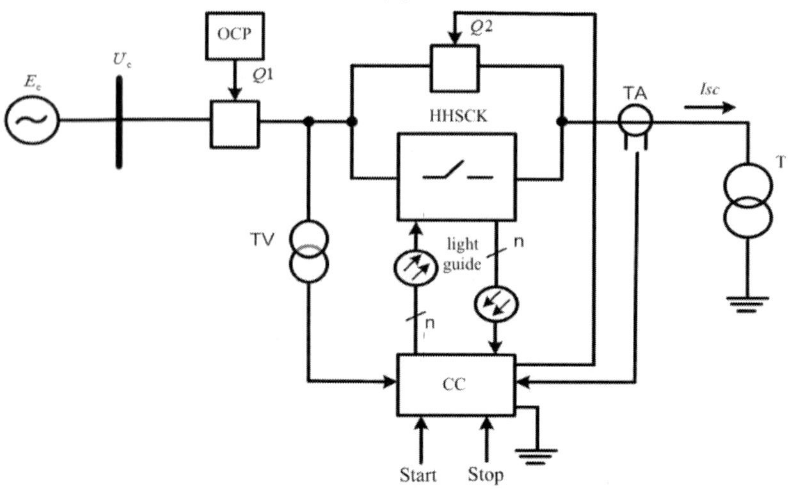

(b)

Figure 7.1. Schemes of inclusion of the HHCSK for the implementation of the complete switching cycle (a) and synchronization of short-circuit experiments (b).

The switch Q1 performs protective functions, providing, upon failure of the HHCSK, the termination of the short-circuit experience as a result of the operation of the overcurrent protection (OCP) with a time delay matched to the required duration of the short-circuit test. Signals of voltage transformers TV and current TA are sent to the control cabinet (CC) for synchronization, respectively, of the turning on and off times of the HHCSK.

In the second case (Figure 7.1b), the HHCSK provides only synchronization of short-circuit tests. From the complete switching cycle, the phase of passing the fault current of the short circuit is excluded. The test circuit is complicated by the introduction of a shunt switch Q2, but the requirements for the switching power of the HHCSK are significantly reduced. The inclusion of the HHCSK must occur at a predetermined phase of the voltage U_C, but with some outstrip of $\Delta t_{os}$with respect to the moment of closing the contacts of the shunt switch Q2. If we provide $\omega \cdot \Delta t_{os} \leq \pi/2$, then the HHSCK will not be affected by the amplitude of the shock current of the short circuit (here ω is the angular frequency of the mains voltage). After closing the contacts, the short-circuit current passes to the circuit of the shunt switch Q2, since the voltage drop across the contact resistance of the contacts is many times less than the voltage drop across the p-n junctions of the PSD.

Switching off of the HHCSK should occur after opening of the contacts of the shunt switch Q2 at the time of the first transition of the short-circuit current through a zero value, i.e., with some delay Δt_{delay}. The length of the delay interval does not exceed the value determined by the inequality $\omega \cdot \Delta t_{delay} \leq \pi$.

Thus, the total duration of the current flow intervals through the HHCSK at the beginning and at the end of the short-circuit test will not exceed one mains voltage period. As a result, the switching power of the synchronizing HHCSK for the test circuit shown in Figure 7.1b, should be substantially less than the switching power of the entire HHCSK of the complete switching cycle for the test circuit shown in Figure 7.1a.

7.2.4. Requirements for Commutation Special Features of Power Semiconductor Devices

Comparative analysis of the switching capacity of various types of PSD is conveniently carried out on the basis of generalized characteristics, which can serve as the nominal switching power

$$S_{SW_{NOM}} = I_{CL} \cdot U_{CL}$$

and the impact shock switching power of a single instrument

$$S_{SW_{IMP}} = I_{IMP} \cdot U_{CL},$$

where I_{CL}, U_{CL} is the classification values of the current and voltage of the PSD; I_{IMP} – impact (impulse) current in the open state.

The nominal switching power $S_{SW_{NOM}}$characterizes the switching capacity of the PSD for long-term operation with the steady-state thermal regime of the semiconductor structure. The impulse switching power $S_{SW_{IMP}}$ characterizes the limiting value of the power that the PSD is able to commute only in a short time, while the temperature regime of the semiconductor structure does not manage to reach thermal equilibrium and exceed the maximum permissible level.

Table 7.1 shows the values of the nominal and impulse switching power of three types of PSD of different manufacturers: open society "Electrovypryamitel" (phototyristor type TF183-2000) [9], ABB Semiconductors (IGCT type 5SHY35L4510) [10] and IXYS (IGBT type T1800GA45A) [11]. Selected PSDs have a tablet construction (the "press-pack" case), the most convenient for manufacturing high-voltage structures, and almost the same diameter of the silicon wafer (≈ 100 mm).

The following designations are adopted in Table 7.1:

U_{DRM}, U_{RRM} – repetitive impulsive direct and reverse voltage of thyristors in the closed state;

U_{CE}, U_R – the maximum allowable DC voltage of the collector-emitter IGBT and constant reverse voltage;

I_{TAV} – average thyristor current in the open state;

I_{TSM} – non-repeating thyristor shock current in the open state;

I_{TQRM} – impulsive lockable current IGCT;

I_C, I_{CM} – the maximum permissible value of the direct current of the collector and the impulsive current of collector IGBT.

Table 7.1. Comparative characteristics of the switching capacity of PSD

Parameter	Device type		
	LTT TF183 – 2000	**IGCT 5SHY35L4510**	**IGBT T1800GA45A**
Classification voltage, kV	U_{DRM} / U_{RRM} 6,5/7,0	U_{DRM} / U_{RRM} 4,5/4,5	U_{CE} / U_R 4,5/0
Classification current, A	I_{TAV} 2115(70°C)	I_{TAV} / I_{TQRM} 1700/4000	I_C 1500
Impackt current, kA	I_{TSM} 40 (10 ms)	I_{TSM} 32 (10 ms)	I_C 3,0 (1,0 ms)
Nominal switching power, MVA	$U_{DRM} \cdot I_{TAV}$ 13,7474	$U_{DRM} \cdot I_{TQRM}$ 18	$U_{CE} \cdot I_C$ 6,75
Shock switching power, MVA	$U_{DRM} \cdot I_{TSM}$ 260	$U_{DRM} \cdot I_{TSM}$ 144	$U_{CE} \cdot I_{CM}$ 13,5

As can be seen from Table 7.1, the value of nominal switching power is quite comparable for the any considered types of PSD. However, the ratio of the impulsive switching power to the nominal

$$K_{SW_{max}} = \frac{S_{SW_{IMP}}}{S_{SW_{NOM}}},$$

which characterizes the limiting overload capacity of the PSD, is for the photothyristors $K_{SW_{max}} \cong 19$, for the lockable thyristors – $K_{SW_{max}} = 8$, for the transistors with the isolated gate only $K_{SW_{max}} = 2$. It should be noted that in the latter case, the permissible duration of the impulse collector current (1 ms) is much less than the permissible duration of the impulse non-repeating current flow through the thyristors (10 ms) [2, 7, 16-23].

Impuls Shock Switching Power of the HHCSK

$$S_{SC_{IMP}} = N \cdot S_{SW_{IMP}} \quad (7.3)$$

is determined by the impulsive switching power of a single PSD and the number N of series-parallel devices connected. The expression (7.3) allows whis taking account (7.2) to determine the minimum number of PSD in the composition of the HHCSK at the nominal parameters of the test transformer

$$N_{min} = \frac{k_{IMP}}{3 \cdot k_{SW_{max}} \cdot S^*_{SW_{NOM}} \cdot (u_{SC\%}/100 + 1/S^*_C)},$$

where $S^*_{SW_{NOM}} = S_{SW_{NOM}}/S_{NOM}$; $S^*_C = S_C/S_{NOM}$.

Since the duration of the short-circuit tests is different from the permissible duration of the impulse shock-current indictated in the information materials, the value of should be chosen so that

$$1 \le k_{SW} \le k_{SW_{max}}.$$

With $k_{SW} = 1$, the the reloading capacity of PSD is practically not used, the number of PSD will be maximal in the composition of the HHCSK. Therefore, it is necessary to take into account the main distinguishing feature of the HHCSK, which consists in the fact that due to the short duration of the short circuit tests of the tested transformer, the temperature regime of the semiconductor structures of the PSD does not reach thermal equilibrium. This allows you to choose $k_{SW} \gg 1$, but not exceeding the limit value $k_{SW_{max}}$. The larger the value of k_{SW}, the less PSD is required for the implementation of the HHCSK, but the temperature regime of semiconductor structures will be more difficult.

The ability of the PSD to withstand the fault current is determined by the maximum permissible transition temperature at the conduction interval, which should not reach values $(400 \div 600)$°C, when processes of lacing forward current begin. The maximum permissible transition temperature at the time of application of the reverse voltage should not reach values (200 ÷ 300)°C due to the process of lashing the reverse current. If a direct voltage is applied to the PSD after the emergency current flows, then the transition temperature should not exceed(150 ÷ 170)°C, since otherwise spontaneous switching [21, 22] is possible.

Therefore, in order to ensure reliable operation of the HHSCK, it is necessary that the temperature of the semiconductor structures of the PSD throughout the entire short-circuit test period, including the voltage restoration phase, does not exceed the allowable value specified in the information materials (usually 120°C or 125°C) [2, 7, 23-30].

Analyzing the classification parameters presented in Table 7.1, it can be concluded that each of the considered PSD can be used to create a HHCSK capable of commuting short-circuit currents of a wide range of power transformers. For power transformers with short-circuit currents on the high-voltage side up to (1,5 ÷ 3,0) kA, there is a principal possibility of creating a complete switching cycle on the basis of high-voltage IGBTs. At high values of the short-circuit current, the HHCSK of the complete switching cycle can be realized on the basis of photothyristors or lockable thyristors IGCT. We should not exclude the variant of synchronizinged HHCSK based on IGBT [2, 7, 16-23].

7.3. Electrical Actions on a High-Voltage Heavy-Current Semiconductor Key during a Short Circuit Test

7.3.1. Characteristics of the Recovering Voltage

To successfully complete the short circuit test, the HHCSK must withstand the application of a recoverable voltage, which is characterized by two parameters - the amplitude value (U_{rec_m}) and the slew rate du_{rec}/dt).

These characteristics of the recovering voltage depend significantly on the dynamic parameters of the PSD, given in Table 7.2.

Table 7.2. Dynamic parameters of PSD

Parametr	Device type		
	LTT TF183 - 2000	**IGCT 5SHY35L4510**	**IGBT T1800GA45A**
Critical rate of voltage rise, B/μS	5000	Does not require restriction	Does not require restriction
Recovery charge, μCoulon	5000	No	No
Switch-on time, μS	t_{don}/t_r 5,0/-	t_{don}/t_r 3,5/1,0	t_{don}/t_r 1,0/2,0
Switch-of time, μS	630	t_{dof}/t_f 7,0/1,0	t_{dof}/t_f 2,0/1,0

Figure 7.2a shows a simplified replacement scheme that allows simulating various modes of switching off the HHCSK depending on the type of PSD used, the presence or absence of an RC damping circuit. Figure 7.2b shows graphs of the change in the short-circuit current in the neighborhood of the zero-point transition point and the recovering voltage $u_B^{(-)}(t)$ for the cases of current termination at the time point t_1, characteristic of IGCT and IGBT and the flow of reverse current in the interval $t_1 - t_3$ typical for thyristors (photothyristors).

The following designations are adopted on the substitution scheme:

- x_C, x_T - equivalent inductive network resistance and inductive component of short-circuit resistance of the tested transformer;
- R_D, C_D - the resistance and capacitance of the RCCP damping RC HHCSK;
- $i_{rr}(t) = I_{rr} \cdot \exp(t/\tau_{rr})$ - current source simulating the flow of a decreasing current in the interval $t_2 - t_3$ of the reverse recovery PSD;

- I_{rr} - amplitude of the reverse current;
- τ_{rr} - a time constant characterizing the rate of reverse current decay;
- U_{CF_m} - amplitude of phase voltage on the busbars of the test bench.

In the general case, the recovering voltage at the HHCSK terminals is of an oscillatory nature and is determined by the expression

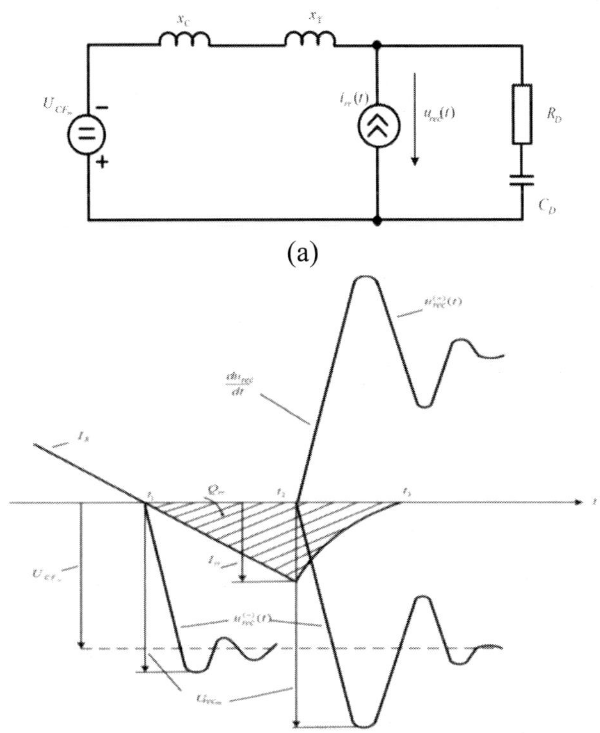

(a)

(b)

Figure 7.2. The replacement circuit (a) for calculating the recovering voltage (b) on the HHCSK after the end of the short circuit test.

$$\begin{aligned} u_{rec}(t) = U_{\mathrm{CF_m}} \cdot [1 + (\sin\omega_B t - \cos\omega_B t)\cdot e^{-\delta t}] + \\ + U_{rr} \cdot [(k_{\delta t} \cdot \sin\omega_B t - \cos\omega_B t)\cdot e^{-\delta t} + e^{-t/\tau_{rr}}] \end{aligned}, \quad (7.4)$$

$$U_{rr} = I_{rr} \cdot R_{\text{D}} \cdot \frac{(\tau_{rr} - \tau_{\text{D}})}{[\tau_{\text{Д}} + 2 \cdot \delta \cdot \tau_{rr} \cdot (\tau_{rr} - \tau_{\text{D}})]}, \tag{7.5}$$

$$k_{\delta r} = \frac{1 + \delta \cdot (\tau_{rr} - \tau_{\text{D}})}{\omega_B \cdot (\tau_{rr} - \tau_{\text{D}})}, \tag{7.6}$$

где $\delta = R_{\text{D}} / 2 \cdot (L_C + L_T)$; $\omega_B = \sqrt{\omega_0^2 - \delta^2}$; $\omega_0 = 1 / \sqrt{(L_C + L_T) \cdot C_{\text{D}}}$;
$L_C = x_C / \omega$; $L_T = x_T / \omega$; $\tau_{\text{Д}} = R_{\text{D}} \cdot C_{\text{D}}$.

For a PSD that conducts a current, the restoring voltage $u_{rec}^{(-)}(t)$ has a negative polarity. This is because the short-circuit current lags by ≈90 g. from the voltage U_C on the busbars of the test bench. For the PSD, which does not conduct current in the last half-cycle of the short-circuit test, the restoring voltage $u_{rec}^{(+)}(t)$ has a positive polarity. If powerful photothyristors are used as a WBS, then the rate of increase of the recovering voltage must not be exceeded, which should not exceed the critical value for this device (see Table 7.2).

Otherwise, it is possible to burn the thyristor structure as a result of uncontrolled inclusion from the effect (du/dt) of the local region of the silicon wafer. With a sequential connection, it is enough to uncontrollably switch on one device to initiate an avalanche-like uncontrolled inclusion of the whole HHCSK.

Expression (7.4) allows us to determine also the rate of increase of the recovering voltage, which takes the maximum value at the initial instant of time

$$\left| \frac{du_{rec}}{dt} \right|_{t=0} = (\omega_B + \delta) \cdot U_{\text{CF}_\text{m}} + (k_{\delta r} \cdot \omega_B + \delta - \frac{1}{\tau_{rr}}) \cdot U_{rr} \tag{7.7}$$

In the absence of a damping circuit ($R_{\mathrm{D}} = 0, C_{\mathrm{D}} = C_B$), the angular frequency of the change in the recovering voltage has a maximum value $\omega_{B_{\max}} = 1/\sqrt{(L_C + L_T)\cdot C_B}$, and expression (7.8) is transformed to the form

$\left|\frac{du_{rec}}{dt}\right|_{t=0} = \omega_{B_{\max}} \cdot U_{\mathrm{CF_m}}$, where C_B is the HHCSK own capacity.

Since $C_B << C_D$, then in the absence of a damping circuit, the rate of increase of the recovering voltage increases substantially and can exceed the critical value many times.

Expressions (7.4) - (7.6) allow, based on the shutdown characteristics ($Q_{rr}, I_{rr}, \tau_{rr}$), to determine the parameters of the damping circuit, which provides the formation of the trajectory of the recovering voltage with the permissible level both of the amplitude (U_{B_m}) and the rate of increase du_B / dt.

In the special case $I_{rr} = 0$, when characteristic of switching off IGCT and IGBT at the time instant (Figure 7.2), expression (7.4) is substantially simplified and takes the form

$$u_{rec}(t) = U_{CF_m} \cdot [1 + (\sin\omega_B t - \cos\omega_B t)\cdot e^{-\delta t}] \quad (7.8)$$

The amplitude of the recovering voltage for IGCT and IGBT can achieve, as seen from (7.8), the limiting value $U_{B_m} = 2 \cdot U_{\mathrm{CF_m}}$ only in the absence of a damping circuit. For LTT, the amplitude of the recovering voltage can take values $U_{B_m} \geq 2 \cdot U_{\mathrm{CF_m}}$ even in the presence of a damping circuit, as can be seen from (7.4), the additional voltage component due to the flow of the reverse current. However, if the synchronization fails, the IGCT or IGBT shutdown will be accompanied by a current cutoff of a magnitude

$$I_{rr} \cong I_{K_m} \cdot (\omega \cdot \Delta t_{\text{error}}),$$

where I_{K_m} is the amplitude of the last half-wave of the short-circuit current; Δt_{error} - sync error.

For example, for $\Delta t_{\text{error}} \approx 100\,\mu S$ a range of short-circuit currents $I_{K_m} = (1 \div 10)kA$, the magnitude of the cutoff current will be very appreciable $I_{rr} \cong (31{,}4 \div 314)A$, quite commensurate with the reverse currents of the thyristors. In this case, the parameters of the recovering voltage will also be determined by the expression (7.4). It should only be accepted $\tau_{rr} \cong 0$, since the current drop interval when IGCT or IGBT is turned off does not exceed 1 μS (see Table 7.2).

Thus, regardless of the type of PSD used, it is necessary to limit either the amplitude of the recovering voltage (IGCT, IGBT) or both the amplitude and the rate of rise of the recovering voltage (LTT). This requires a damping RC-chain, which is most rational to introduce into the HHCSK. In this case, the problem of evenly dividing the voltage between the PSD in the closed state and at the switching intervals is additionally solved [7, 16, 19, 31-33].

7.3.2. High-Voltage Heavy Current Semiconductor Key with Using Photothyristors

VTSK on photothyristors. Figure 7.3 shows the power circuit of the HHCSK, implemented on photothyristors $VS_1^{(+)},\ldots,VS_n^{(+)}$, forming a branch for the flow of a positive half-wave of the short-circuit current, photo-thyristors $VS_1^{(-)},\ldots,VS_n^{(-)}$ forming a branch for the flow of a negative half-wave of the short-circuit current, and a common damping circuit $R_1C_1,\ldots,R_nC_n$ that also provides a uniform voltage division.

It is advisable to use photothyristors of the TF183-2000 type, which have a high ($k_{\Pi_{MAX}} = 19$) overload capacity. As can be seen from (7.4), this makes it possible to avoid parallel connection of devices and to minimize the number of photothyristors in the composition of a HHCSK capable of carrying out a full switching cycle.

To turn on the photothyristor TF183 - 2000, an optical signal with a power of 40 mW and a duration of 10 μS is sufficient. In this case, the delay time for switching on the photothyristor does not exceed 5 μS (Table 7.2). The control pulse is transmitted to the photosensitive region of the silicon structure through an optical fiber, which is connected optically by one end to the photothyristor housing and the other end to a laser diode in the CU.

For simultaneous connection of series-connected photothyristors, it is advisable to use a serial connection of laser diodes connected to a common electrical driver. Attenuation of the optical signal in the optical fiber is very small, so the length of the light guide can be chosen only by the condition of providing the necessary high-voltage galvanic isolation between the power circuit HHCSK and the control cabinet.

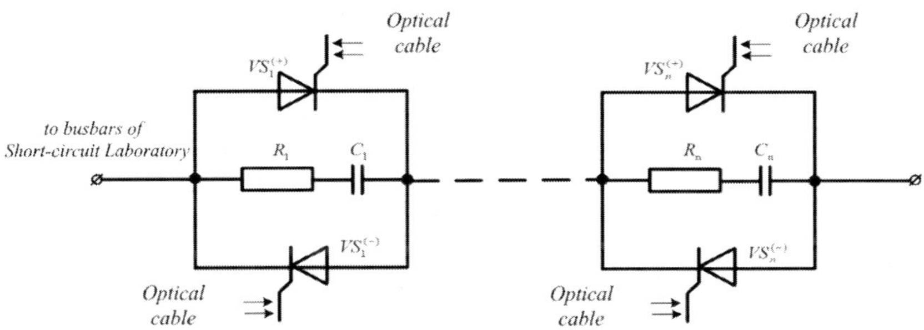

Figure 7.3. Power circuit of the HHCSK on LTT photothyristors.

The silicon structure of the TF183-2000 phototyristor is integrated with protection against breakdown at direct overvoltages, which forms an internal control signal and includes a photothyristor with an increase in the forward voltage to the level of 6500 V (Table 7.1). Integrated protection against breakdown allows almost completely to abandon the electronic components in the control channel located at a high potential, and to simplify the power scheme of the HHCSK as much as possible [31-36].

7.3.3. Emergency Interruption of the Short Circuit Testing by a Photothyristor High-Voltage Heavy Current Semiconductor Key

Figure 7.4 shows a simplified scheme of electrodynamic tests of an power transformer TT in a two-winding mode with an artificial short-circuit of the terminals on the low-voltage side (a) and time diagrams of voltage and current in the short-circuit test with the action of the largest aperiodic component (b). Until then t_1, the voltage U_C of the busbars of the test bench is applied to the photothyristors $VS1_{(1)},\ldots VS1_{(n)}$ and $VS2_{(1)},\ldots VS2_{(n)}$.

At the time t_1 when a positive voltage appears on the photothyristors $VS1_{(1)},\ldots VS1_{(n)}$, a short-circuit experience with the angle of the initial inclusion of HHCSK the equal to zero begins.

A short-circuit current flows through the HHCSK, containing periodic $i_{\mathrm{sc(p)}}(t)$ and aperiodic $i_{\mathrm{sc(ac}}(t)$ (see Figure 7.4b) components

$$i_{\mathrm{sc}}(t) = I_{\kappa_m} \cdot \left[e^{-t/\tau} - \cos\omega t \right], \tag{7.9}$$

where $I_{sc_m} = \sqrt{2} \cdot E_C / Z_\Sigma$ is the amplitude of the steady short-circuit current on the high-voltage side of the power transformer under test; E_C - effective value of the phase EMF of the electrical network supplying the test network;

- $Z_\Sigma = \sqrt{(x_C + x_T)^2 + r_\mathrm{T}^2}$ module of impedance of short circuit; x_C - equivalent inductive resistance of the electrical network supplying the test bed; x_T, r_T - inductive and active components of short-circuit resistance of the power transformer under test;

$\tau = (x_C + x_T)/\omega \cdot r_T$ - time constant of the short circuit; ω -angular frequency of mains voltage.

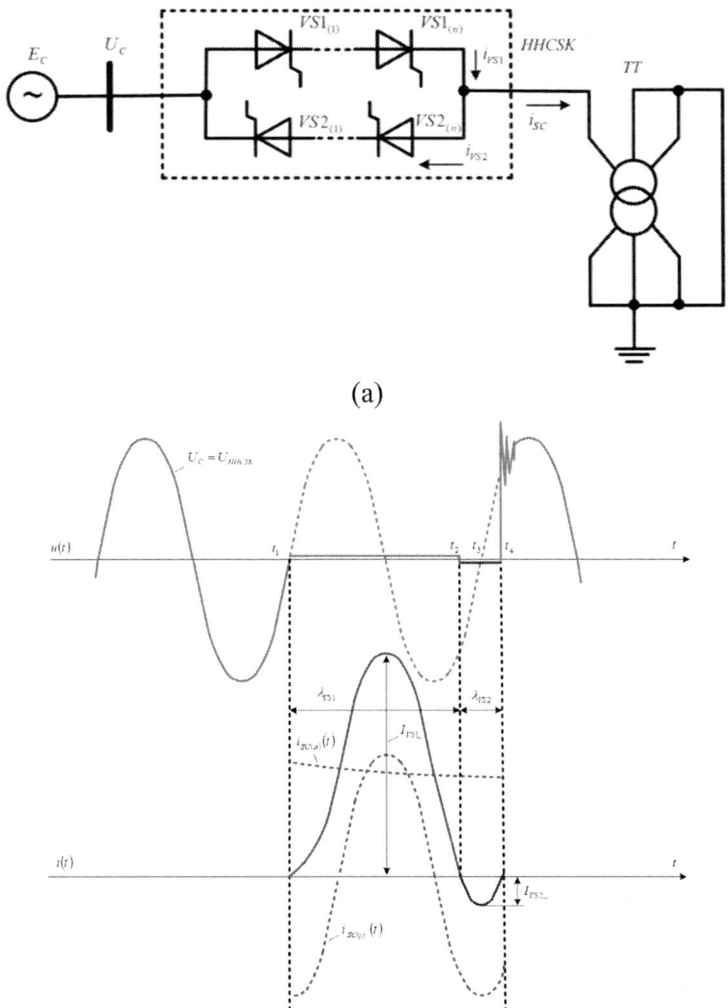

(a)

(b)

Figure 7.4. Simplified circuit of electrodynamic tests (a) and time diagrams of voltages and currents of the short-circuit test (b).

In the first period of the short circuit test of the winding of the power transformer under test, the TT and photothyristors $VS1_{(1)},\ldots VS1_{(n)}$ are subjected to shock current with amplitude

$$I_{VS1_m}^{(1)} = I_{\text{к}_\text{m}} \cdot \left(1 + e^{-1/2\tau^*}\right), \tag{7.10}$$

where, $\tau^* = \tau/T$, T is the period of the mains voltage.

The duration of the conductivity interval $(t_1 - t_2)$ of the photothyristors $VS1_{(1)},\ldots VS1_{(n)}$ and, respectively, the short-circuit shock current will be

$$\lambda_{VS1} = 2\pi - \arccos\left(e^{-1/\tau^*} \right). \tag{7.11}$$

At the time t_2, photothyristors $VS2_{(1)},\ldots VS2_{(n)}$, are switched on, through which a short-circuit current with amplitude

$$I_{VS2_m} = I_{\text{к}_\text{m}} \cdot \left(1 - e^{-1/\tau^*}\right) \tag{7.12}$$

and duration

$$\lambda_{VS2} = 2 \cdot \arccos\left(e^{-1/\tau^*} \right). \tag{7.13}$$

For example, for $\tau^* = 9$ the amplitudes of the positive (7.10) and negative (7.12) half-waves of the short-circuit current are equal to $I_{VS1_m} = 1{,}95 \cdot I_{\text{SC}_\text{m}}$ and $I_{VS2_m} = 0{,}11 \cdot I_{\text{SC}_\text{m}}$. The conductivity intervals (7.11) of the photothyristors $VS1_{(1)},\ldots VS1_{(n)}$ and (7.14) of the photothyristors

$VS2_{(1)},\ldots VS2_{(n)}$ will be $\lambda_{VS1} \cong 333{,}5$ el. angle and $\lambda_{VS2} \cong 53$ el. angle respectively.

On the interval $(t_1 - t_2)$ when the shock current of the short-circuit of the emergency processes occurs in the windings of the power transformer under test, the electrodynamic tests can be interrupted either at the time t_2 or at the time t_4.

In the first case, the reverse voltage will be applied to the photothyristors $VS1_{(1)},\ldots VS1_{(n)}$ up to the time point t_3, and from the moment of time t_3 the photothyristors $VS1_{(1)},\ldots VS1_{(n)}$ must withstand the application of direct voltage.

For this it is necessary that the turn-off time of the photothyristors $VS1_{(1)},\ldots VS1_{(n)}$ satisfy the condition

$$t_q < \left[\left(\frac{2\pi - \lambda_{\text{VS1}}}{\omega}\right) - t_{rr}\right], \tag{7.14}$$

at an aperiodic form of the recovering stress or condition

$$t_q < 1/\text{f}_{\text{recovery.}} \tag{7.15}$$

with the vibrational form of the recovering voltage.

(Here t_{rr} -- the time of the reverse restoration of the power thyristors, f_{recovery} - the frequency of the recovering voltage).

Violation of the conditions (7.14) or (7.15) will lead to uncontrolled (without the supply of control pulses) the inclusion of the HHCSK at the time and the subsequent irreversible breakdown of all photothyristors $VS1_{(1)},\ldots VS1_{(n)}$.

In the second case, an inverse voltage equal to the voltage drop across the photothyristors $VS2_{(1)},\ldots VS2_{(n)}$ is applied to the photothyristors $VS1_{(1)},\ldots VS1_{(n)}$ in the interval $(t_2 - t_4)$. The magnitude of the reverse voltage on the photothyristor $VS1_{(i)}$, equal to the voltage drop across the phototthyristor $VS2_{(i)}$, does not exceed (1 2) V. Thus, the restoration of the locking properties of the photothyristors $VS1_{(1)},\ldots VS1_{(n)}$ will occur practically without applying an inverse voltage. As a result, the turn-off time increases by (1.5 2) times and must satisfy the condition

$$t_q < \frac{\lambda_{VS2}}{(1{,}5 \div 2{,}0)\cdot\omega}, \tag{7.16}$$

violation of which will also lead to uncontrolled inclusion of photothyristors $VS1_{(1)},\ldots VS1_{(n)}$.

The fulfillment of the conditions (7.14) ÷ (7.16) depends on the parameters of the tested transformer (x_T, r_T), the damping circuit parameters of the HHCSK (in Figure 7.4, not shown), the characteristics of the switching off $(t_q, t_{rr})$of the photothyristors. It should be added that the turn-off time of the powerful low-frequency photothyristors that can be used in the HHCSK reaches values (600 ÷800) µS. This is quite commensurate with the time intervals $(t_2 - t_3)$ and $(t_2 - t_4)$, shown in Figure 7.4, b.

If the conditions (7.14) (7.16) are not satisfied, then by means of (7.9) it is necessary to determine the number of periods after which an interruption of the short-circuit experiment is possible. Accordingly, it is determined that the minimum number of periods of the short-circuit experience during which the windings of the power transformer, even in the presence of emergency deformations, will be subjected to electrodynamic influences. Switching devices based on power thyristors with electric and direct light control are theoretically able to interrupt the short-circuit experience in a time interval of no more than 10 mS after the emergence of an emergency process [7, 16,

22-25]. However, this is only possible if the short-circuit current does not contain an aperiodic component. Otherwise, the emergency termination of the short-circuit test, for example, in the first period, when the amplitude of the shock current and the likelihood of the development of the emergency process, are maximum, may result in subsequent uncontrolled power thyristors switching and irreversible damage to semiconductor structures. This does not allow to fully ensure the safe conduct of electrodynamic tests.

7.4. The Technology of Safe Electrodynamic Testing

7.4.1. Scheme of Safe Conduct of Electrodynamic Tests

The technological process of electrodynamic testing of power transformers in accordance with the requirements of GOST 20243-88 provides for carrying out control measurements before and after the end of each short-circuit test [1-7]. This is necessary to assess the initial state of the power transformer under test, to detect initial damage and to monitor the development of damage during the test. Among the most informative diagnostic parameters is the inductive short-circuit resistance, the change of which allows fixing axial and radial deformations of the windings, as well as changes in the mutual spatial arrangement of the windings of the power transformer under test [7, 25].

For the practical implementation of safe electrodynamic testing technology, monitoring of the diagnostic parameters of a power transformer is necessary not only before and after each short circuit test, but also during a short circuit test and, accordingly, under operating voltage. This allows to fix the beginning of the occurrence of the emergency process and to reduce the scale of the accident in the event of a timely disconnection of the test transformer. Information-measuring systems (IMS) for monitoring the parameters of the tested transformer during short-circuit tests have been considered in a number of works [7, 16, 17, 25].

However, only the formation of an emergency stop signal of a short-circuit test is not enough to ensure the safety of electrodynamic tests of power transformers. It is necessary to minimize the time interval between the moment of formation of the IIS of the alarm signal and the moment of emergency termination of the short-circuit experiment by forcibly interrupting the short-circuit current. Only in this case it is possible to prevent the accidental destruction of the windings of the power transformer under test and thereby increase the safety of the electrodynamic tests.

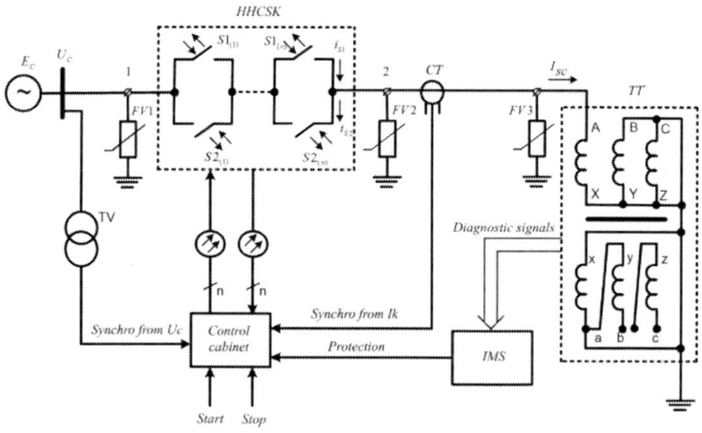

Figure 7.5. The scheme of safe conduct of electrodynamic tests with the control of the condition of the tested power transformer.

Figure 7.5 presents a simplified block diagram allowing to ensure the safety of the short circuit tests of the TT power transformer under test by rapid interruption of the short-circuit current with the HHCSK

The power circuit of the HHCSK contains a chain of n consecutively connected cells, each of which is formed by oppositely-connected keys $S1_{(1)} - S2_{(1)} \ldots S1_{(n)} - S2_{(n)}$ .. As the latter, should be used fully controlled power semiconductor devices, for example, lockable thyristors with a ring lead of the control electrode and an integrated control device (IGCT) or powerful bipolar transistors with an isolated gate (IGBT).

The control cabinet (CC), on the "Start" signal, forms a series of control pulses, which are transmitted through the light guides to the HHCSK potential. The number of control pulses determines the number of positive i_{S1} and negative i_{S2} half-periods of the current in the short-circuit test, and the initial phase of the first control pulse of the relative mains voltage determines the amount of the aperiodic component of the short-circuit current. At the stop signal, the series of control pulses stops, completing the short-circuit test. The signals of the voltage transformers TV and TA current ensure synchronization of the switching and deactivation times of the HHCSK, respectively. In this case, the normal completion of the short-circuit test involves turning off the HHCSK only at the moment of transition I_K through the zero value.

The Information and Measuring System (IMS) processes the diagnostic signals received from the primary measuring transducers (not shown in Figure 7.5) that monitor the current state of the TT during the short-circuit test, for example, the average value of the short-circuit inductance over a period. In case of a dangerous deviation of the monitored parameter from the initial value of the IIS, a "Protection" signal is generated, which is transmitted to the CC. The latter forms a signal to lock the power semiconductor devices HHCSK regardless of the current value of the short-circuit current, carrying out an emergency termination of the short-circuit test [26-30].

It must be borne in mind that the emergency termination of the short-circuit test not synchronized with the moment of the current I_{SC} passing through the zero value will be accompanied by dangerous overvoltages both on the high voltage winding (A-X) of the TT and at the HHCSK terminals. The multiplicity of overvoltages will be determined by the amount of energy accumulated by the moment of switching off the HHCSK in the magnetic field of the short-circuit inductance, the inductances of the current leads between the busbars of the test bed and the terminal "1" of the HHCSK, between the terminal "2" of the HHCSK and the terminal "A" of the high voltage winding TT. To absorb the energy of the magnetic field and to reduce the level of overvoltages in case of emergency interruption of the short-circuit test, the surge arrester $FV1$ and $FV3$ are used in Short-circuit Laboratory scheme.

The limiter $FV3$ absorbs the energy of the magnetic field of the short-circuit inductance of the TT and protects from overvoltage the high-voltage winding A-X. The limiter absorbs the energy of the magnetic field of the current conductor between the terminal "2" of the HHCSK and the terminal "A" of the TT and fixes the potential of the output 2 relative to the "ground". The limiter $FV2$ absorbs the energy of the magnetic field of the internal HHCSK bus and fixes the potential of output 1 relative to the "ground". In addition, the limiter $FV1$ protects the HHCSK from lightning and switching overvoltages coming to the busbars U_C of the test stand from the supply network side.

The most severe impact in the emergency interruption of the short-circuit test is an overvoltage limiter $FV3$, since the short-circuit inductance of the TT system is many times greater than the inductance of the test-bed. The most unfavorable for the emergency interruption of the short circuit with an aperiodic component is the moment of the peak of the shock current in the first half-period (see Figure 7.4b), determined by the expression (7.10).

In this case, varistors of the overvoltage limiter $FV3$ after an emergency shutdown of the HHCSK must absorb shock energy W_{SE} whose value is equal to

$$W_{\mathrm{SE}} = \frac{L_K \cdot (I^{(1)}_{\mathrm{sc_m}})^2}{2} = \frac{u_{K\%}}{100} \cdot K_{ap}^2 \cdot \left(\frac{I_{sc}}{I_{nom}} \right)^2 \cdot \frac{S_{nom}}{3 \cdot \omega}, \tag{7.17}$$

where S_{NOM}, I_{NOM} - rated power and rated current of testing transformer; $L_K, u_{K\%}$ - inductance and short-circuit voltage of tested transformer; $K_{ap.}$ - is the the value of the impact coefficient (aperiodic coefficient) of short-circuit current, the value of which for powerful transformers can reach values (1.8 - 1.95).

7.4.2. The Energy of Testing Transformer When the Short-Circuit Test Fails

Taking into account (7.1) and taking as energy the basis energy

$$W_{\delta} = (S_{nom} \cdot T),$$

the expression (7.17) can be transformed to a form allowing to determine the magnitude of the shock energy W_{SE} from the passport data of the power transformer under test

$$W_{se}^{*} = \frac{W_{se}}{W_{б}} = \frac{K_{ap}^{2}}{6\pi} \cdot \frac{(u_{K\%}/100)}{\left[(1/S_{SC}^{*}) + (u_{K\%}/100)\right]^{2}}, \tag{7.18}$$

where; $S_{C}^{*} = S_{C}/S_{NOM}$; Ssc - short-circuit capasity of supply system on the busbars of Short-circuit Laboratory.

The duration of the process of absorbing the energy of the TT transformer under test in case of an abrupt termination of the short-circuit test can be estimated assuming that the remaining voltage U_{rem} of the overvoltage limiter $FV3$ remains unchanged

$$\Delta t \cong \frac{L_{K} \cdot I_{ap_{\max}}}{U_{\text{rem}}} = \frac{x_{K} \cdot I_{\text{ap}_{\max}}}{k_{\text{pi}} \cdot \omega \cdot U_{nom}},$$

where L_{K} is the short-circuit inductance of the transformer under test; $k_{pi} = (1{,}6 \div 1{,}8)$ - protective index of nonlinear overvoltage suppressor; U_{NOM} - rated voltage of the high-voltage winding of the transformer TT under test.

For quantitative estimates, it is much more convenient to use the relative duration of the considered interval

$$\Delta t^* = \frac{\Delta t}{T} = \frac{1}{\pi \cdot \sqrt{6}} \cdot \frac{k_{ap}}{k_{pi}} \cdot \frac{(u_{K\%}/100)}{\left[(u_{K\%}/100) + (1/S_C^*)\right]}, \tag{7.19}$$

which can easily be determined from the passport data of the transformer under test. For example, in case of emergency interruption of the short-circuit test of a power transformer with a rated power $S_{NOM} = 63\ MVA$, Vthe relative values of the shock energy and the duration of the process of absorbing the stored energy will be $W_{SE}^* \cong 1{,}375$, $\Delta t^* \cong 0{,}13$, (in named units $W_{SE} \cong 1732{,}5 kJ$, $\Delta t \cong 2{,}6 mS$, respectively).

Expressions (7.10), (7.18) and (7.19) make it possible to determine the level of electrical influences on the overvoltage limiter $FV3$ in the event of an emergency termination of the short-circuit test. If it is necessary to increase the capacity of the surge suppressor $FV3$, parallel varistor columns or the parallel connection of several protective devices should be used. Similarly, using the results of measuring the inductance of busbars of Short-circuit Laboratory and HHCSK, it is possible to estimate the requirements for the bandwidth of the surge arresters $FV1$ and $FV2$.

7.5. Schemes of High-Voltage High-Current Semiconductor Switches on Fully Controlled Power Semiconductor Devices

7.5.1. High-Voltage High-Current Semiconductor Keys on Lockable Thyristors IGCT

HHCSK on lockable thyristors IGCT. Lockable thyristors with an integrated Gate-Commutated Thyristor (IGCT) have been developed relatively recently based on conventional gate turn-off (GTO) thyristors [7, 22, 31].

IGCTs have a tablet design (a "press pack" casing) with a ring lead of the control electrode, which surrounds the device in a circle and passes through the ceramic body, contacting the outside with the driver. This made it possible to radically decrease the inductance of the control electrode circuit and achieve the rate of reduction of the switched-off current to ≈ 4 kA/μS, and the shutdown time was reduced to microseconds (Table 7.2). Short shutdown time IGCT allows you to synchronize the turn-off time of the HHCSK with an accuracy of several microseconds, providing a non-commutated switching and, accordingly, the end of the short-circuit experience with a minimum amplitude of the recovering voltage. For comparison, the turn-off time of the TF183-2000 phototyristor is 630 μS.

It should be noted that the rather wide nomenclature of IGCT is focused primarily on the use of high voltage voltage inverters in the structure and is structurally implemented in general with a built-in reverse diode in the "press-pack" package, i.e., does not have an inverse blocking ability. However, such IGCTs can also be used to create a HHCSK.

Figure 7.6 shows two variants of the HHCSK power circuit, one on IGCT having reverse blocking capacity (a), the other on IGCT, which do not have the reverse blocking capacity (b). In the first case, each thyristor HHCSK cell is formed by the counter-parallel connection of two IGCTs, in the second case each thyristor HHCSK cell is formed by the counter-series connection of two IGCTs.

The power circuit of the HHCSK, shown in Figure 7.6a, is implemented on lockable thyristors $VS_1^{(+)},...,VS_n^{(+)}$ and $VS_1^{(-)},...,VS_n^{(-)}$ forming branches for the flow of positive and negative half-waves of the short-circuit current, respectively. The dashed lines in Figure 7.6a show unified structural units: a lockable thyristor $VS_n^{(+)}$ -driver $CU_n^{(+)}$.

Control signals are transmitted to the drivers $CU_1^{(+)},...,CU_n^{(+)}$ and $CU_1^{(-)},...,CU_n^{(-)}$ via fiber-optic cables from the ground potential. The drivers $CU_1^{(+)},...,CU_n^{(+)}, U_1^{(-)},...,U_n^{(-)}$ are powered by a constant voltage (usually 20 V) from external power supplies and correspondingly up to several tens of watts each.

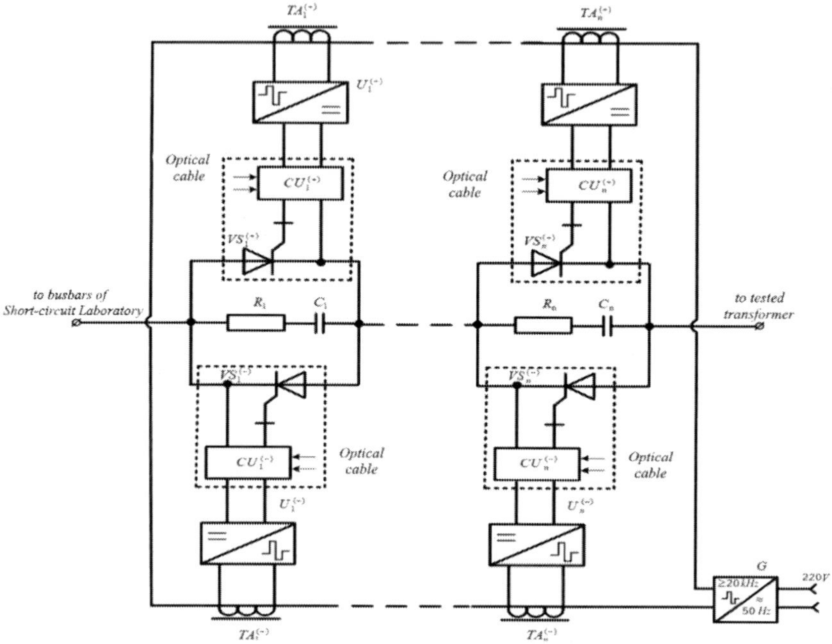

(a)

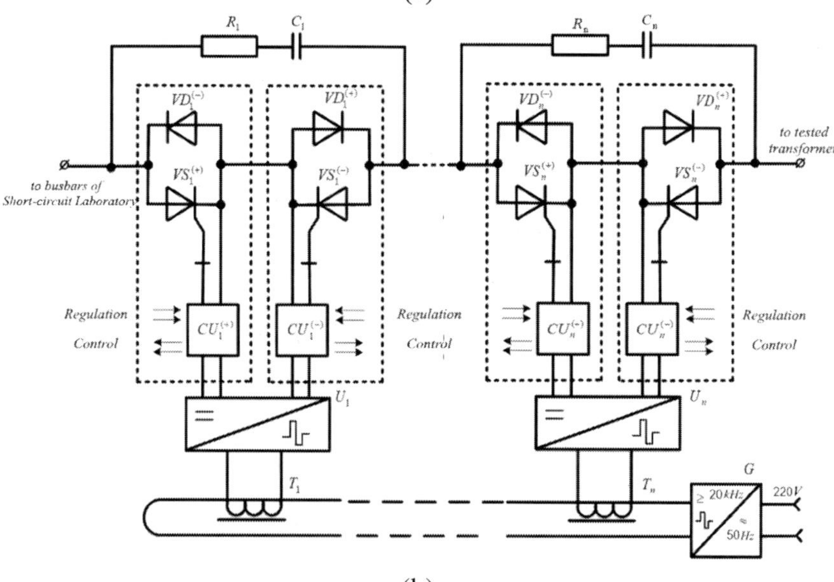

(b)

Figure 7.6. Variants of the power scheme HHCSK for IGCT with (a) and not (b) reverse blocking capacity with elements of a high-potential control system.

Galvanic isolation of the voltage converters $U_1^{(+)},...,U_n^{(+)}$ and $U_1^{(-)},...,U_n^{(-)}$, each of which is located at the potential of the cathodes of the corresponding lockable thyristors $VS_1^{(+)},...,VS_n^{(+)}$ and $VS_1^{(-)},...,VS_n^{(-)}$, is provided by the cable and transformer channel $TA_1^{(+)},...,TA_n^{(+)}$ - $TA_1^{(-)},...,TA_n^{(-)}$.

The primary winding of the cable and transformer channel is common for all transformers $TA_1^{(+)},...,TA_n^{(+)}$ and $TA_1^{(-)},...,TA_n^{(-)}$, secondary windings are connected to power supplies $U_1^{(+)},...,U_n^{(+)}$ and $U_1^{(-)},...,U_n^{(-)}$, accordingly. The high-frequency voltage generator G feeds the primary winding of the cable-transformer channel.

The assignment of the main elements of the power circuit of the HHCSK, shown in Figure 7.6b, does not require additional explanations. It should be added that a positive half-wave of the short-circuit current flows through thyristors $VS_1^{(+)},...,VS_n^{(+)}$ and diodes $VD_1^{(+)},...,VD_n^{(+)}$, and the negative half-wave through thyristors $VS_1^{(-)},...,VS_n^{(-)}$ and diodes $VD_1^{(-)},...,VD_n^{(-)}$. As a result, the power losses in each IGCT package (indicated by the dotted lines in Figure 7.6b are composed of losses in the thyristor and losses in the diode, which significantly increases the temperature regime of the semiconductor structures. Therefore, the shock switching power of the HHCSK, realized according to the variant of Figure 7.6b, is fundamentally less than the shock switching power of the HHCSK realized in the variant of Figure 7.6a. In general, however, the lockable IGCT thyristors with an equal area of the silicon wafer are inferior to the photo-thyristors by the magnitude of the shock switching power (almost 2.4 times). Therefore, all other things being equal, a much larger number of locked thyristors of IGCT (also $\approx$ 2.4 times) will be required to realize the HHCSK. And this despite the fact that the power circuit HHCSK of the IGCT is much more complicated than that of the photothyristors due to the large number of electronic components of the control channel at high potential.

Nevertheless, an IGCT can be HHCSK implemented with a full switching cycle, capable of providing accurate synchronization for the non-existent end of the short-circuit test. In the event of an emergency, the IGCT-HHCSK is able to interrupt the short-circuit experiment and at a non-zero current [31-36].

7.5.2. High-Voltage High-Current Semiconductor Keys on Bipolar Transistors with Isolated Gate IGBT

It should be noted that bipolar transistors in principle are not capable of blocking the reverse voltage. Therefore, single IGBTs with a reverse diode, structurally placed in a common "press-pack" type enclosure, are most convenient for use in the HHCSK structure. The series connection of two IGBT allows to receive the AC key.

The press-pack of the IGBT type T1800GA45A of the manufacturer IXYS can be used in the HHCSK, the main technical characteristics of which are given in Tables 7.1 and 7.2. To control the IGBT, a driver C0030BG400SaS with a supply voltage of ± 15 V is provided, providing a pulse of 30 A output current with a rise time of 0.5 μS when operating on a capacitive load of 100 nF (gate-to-emitter capacitance of the T1800GA45A transistor).

Figure 7.7 shows the power circuit of the HHCSK, which consists of identical series-connected AC switches.

Each AC key is formed by the counter-sequential inclusion of two Press-pack IGBTs with reverse diodes: $VT_1^{(+)} - VD_1^{(-)}$ and $VT_1^{(-)} - VD_1^{(+)}$, ..., $VT_n^{(+)} - VD_n^{(-)}$ and $VT_n^{(-)} - VD_n^{(+)}$.. Each device "press-pack" is marked with dotted lines.

A positive half-wave of the short-circuit current flows through the transistors $VT_1^{(+)},...,VT_n^{(+)}$ and the reverse diodes $VD_1^{(+)},...,VD_n^{(+)}$. A negative half-wave of the short-circuit current flows through the transistors $VT_1^{(-)},...,VT_n^{(-)}$ and reverse $VD_1^{(-)},...,VD_n^{(-)}$ diodes.

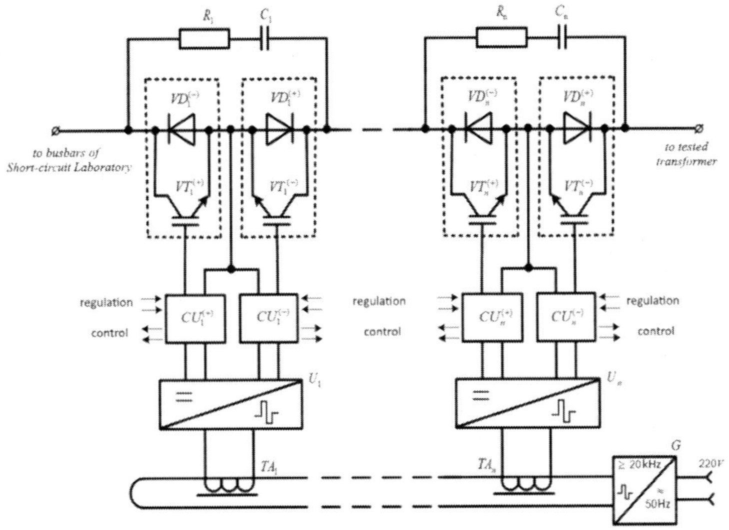

Figure 7.7. Power circuit HHCSK on bipolar transistors with isolated gate (IGBT) with elements of the control system at high potential.

Each transistor $VT_1^{(+)},...,VT_n^{(+)}$ and $VT_1^{(-)},...,VT_n^{(-)}$ is controlled by an individual driver $CU_1^{(+)},...,CU_n^{(+)}$ and $CU_1^{(-)},...,CU_n^{(-)}$, accordingly,. In this case, the pair of drivers $CU_1^{(+)}$ and $CU_1^{(-)}$,..., $CU_n^{(+)}$ and $CU_n^{(-)}$ are on the same potential, allowing to provide power to each pair of drivers from one source of stabilized bipolar voltage $U_1,..U_n$, respectively. The input for each power source $U_1,..U_n$ is the high-frequency alternating voltage of the secondary windings of transformers $TA_1,...,TA_n$ with a common primary winding connected to the high-frequency generator G. The cable-transformer channel $TA_1,...,TA_n$ provides the transfer of the high-frequency voltage of the generator G from the ground potential to the potentials of the series-connected HHCSK cells. The isolation level between the primary winding and the secondary windings of the transformers $TA_1,...,TA_n$ must withstand the total voltage of the HHCSK.

Given that the IGBT's overload capability is very small compared to thyristors (usually $I_{CM} \leq 2 \cdot I_C$ when the collector current pulse duration is not more than 1 mS), the most preferred use of the HHCSK with the power circuit shown in Figure 7.7 is synchronization of the moments of the beginning and end of the short-circuit tests (cm see Figure 7.1b). The control algorithm of the HHCSK in this case must provide for supplying from the ground potential to all drivers $CU_1^{(+)}, CU_1^{(-)},..., CU_n^{(+)}, CU_n^{(-)}$ simultaneously a "wide" light pulse whose beginning is synchronized with the voltage on the busbars of the test bed, and the end is synchronized with the instant of the short-circuit current passing through zero. In addition, the light pulse must be synchronized with the commands for switching on and off the Q2 switch (see Figure 7.1b) in such a way that the collector current of the IGBT does not exceed the allowable value (for example, for transistors of type T1800GA45A this is a pulse collector current of 3000 A for 1 mS).

When testing power transformers with the maximum amplitude of the short-circuit shock current not exceeding 3 kA, the HHCSK with the power circuit on the IGBT can be switched on according to the scheme shown in Figure 7.1a, i.e., used to implement a complete switching cycle [7, 22, 31, 36-49].

Conclusion

1. A complex of functional requirements for switching equipment for electrodynamic testing of power transformers, including pre-magnetization of the magnetic circuit, synchronization, regulation of the initial phase of the inclusion, transmission and disconnection of the short-circuit current, can be performed in the fullest possible only with a high-voltage high-current semiconductor switch.
2. For the comparative characterization of various versions of the HHCSK implementation, it is advisable to use the concepts of shock and nominal switching power, the value of which for HHCSK is generally determined by the tested transformer and the adjacent

network, and for individual PSD it is determined by the classification parameters indicated in the information materials.

3. Depending on the magnitude of the shock switching power, the HHCSK during electrodynamic testing can be used either alone to implement a full switching cycle, including synchronization and short-circuit current transmission, or together with a shunt switch only to synchronize short-circuit experiments. during electrodynamic testing can be used either alone to implement a full switching cycle, including synchronization and short-circuit current transmission, or together with a shunt switch only to synchronize short-circuit experiments.
4. To complete the short circuit test, it is necessary to form a trajectory of the voltage restored to the HHCSK for any type of PSD used, taking into account the amplitude and growth rate limitations with the damping RC circuit introduced into the HHCSK.
5. HHCSK with a maximum value of shock switching power with a minimum amount of PSD and the most simple power circuit is realized on the basis of photothyristors.
6. The photothyristor HHCSK is not able to provide emergency termination of electrodynamic tests during the first period of the short-circuit experiment with the aperiodic component of the current, when the maximum probability of the development of accidental processes in the windings is due to the impact of the shock current, since for low-frequency power thyristors, due to asymmetry in the conductivity intervals, application of reverse voltage to restore locking properties.
7. The safety of the electrodynamic tests of power transformers is ensured by the forced termination of the short-circuit test immediately after the formation of the emergency condition signal of the winding of the power transformer under test at an arbitrary value of the short-circuit current and can be practically realized only with the help of HHCSK on fully controlled power semiconductor devices - IGCT or IGBT.

8. An abnormal termination (in a few microseconds) of the short-circuit test is possible if the total energy absorbed in the magnetic field of the short-circuit inductance and the test bed are ensured, for example, by means of non-linear surge arresters installed on the terminals of both the HHCSK and high-voltage windings voltage of the transformer under test.
9. The non-linear surge suppressor installed on the high voltage winding of the tested transformer undergoes the most severe impacts, since in the case of termination of the short-circuit test at the time of the shock current, it must absorb all the energy of the magnetic field accumulated in the short-circuit inductor of the transformer under test.
10. HHCSK based IGCT or IGBT have additional functionality to stop almost instantaneous short-circuit experience at any time in the event of an emergency, but is much more difficult to implement due to the large number of PSD and electronic components of the control channel at high potential of the power circuit.

Acknowledgments

The authors express deep gratitude to Dr. Professor, Honored Scientist of the Russian Federation, full member of the Academy of Electrotechnical Sciences Shakaryan Yu. G. for a number of valuable comments, which allowed to significantly expand the scope of the study.

References

[1] Khrennikov, A.Yu., Kuvshinov, A.A., Mazurov, V.A., Nesterchuk, A.K. (2015). Control of Power Transformer Windings during Electrodynamic Test Using High-Voltage Thyristor Valve for Regulation the Angle of the Initial Inclusion. *International Journal of*

Automation and Control Engineering, Vol. 4, No. 2, 58-66. www.sei pub.org/ijace.

[2] Khrennikov, A.Yu.; Shakaryan, Y.G.; Dementyev, Y.A. (2013). Shortcurrent Testing Laboratories. Short-Circuit Performance of Power Transformers, Transformer Testing Experience. *International Journal of Automation and Control Engineering (IJACE)*, Volume 2, Issue 3. (pp. 120-127). http://www.seipub.org/ijace/ AllIssues.aspx.

[3] Khrennikov, A.Yu. (2004). Short-circuit performance of power transformers. Test experience, including fault diagnostic, *Reports of Advanced Research Workshop on Modern Transformers* (ARWtr 2004), Vigo – Spain.

[4] Khrennikov, A.Yu. (2006). Diagnostics of electrical equipment's faults, defects and weaknesses, Reports of *Conference on Condition Monitoring and Diagnosis (CMD 2006)*, Korea.

[5] Portales, E.; Filion, Y.; Mercier, A. (2011). Transformer-controlled switching taking into account residual flux-modelling transformers for realistic results in system studies, *RECIFE*, 68.

[6] Bertagnolli, G. (1998). Short-Circuit Duty of Power Transformers. *ABB Transformatori*, Legnano (Milano) – Italy., 197.

[7] Khrennikov, A.Yu. (2016). *Diagnostics of Electrical Equipment Faults and Power Overhead Transmission Line Condition by Monitoring Systems (Smart Grid): Short-Circuit Testing of Power Transformers*, Nova Publishers, New York, USA, 165. https://www. novapublishers. com/catalog/product_info.php?products_id=56561.

[8] Khrennikov, A.Yu. (2012). Monitoring information-measuring system for detecting power transformer faults, FRA and LVI-testing diagnostics experience, *Reports of Asia-Pacific Power and Energy Engineering Conference (APPEEC 2012)*, Shanghai, China.

[9] Khrennikov, A.Yu. (2012). New "intellectual networks" (Smart Grid) for detecting electrical equipment faults, defects and weaknesses. *Smart Grid and Renewable Energy*, Volume 3, Issue 1.

[10] Khrennikov, A.Yu. (2013). Smart Grid technologies for Detecting Electrical Equipment Faults, Defects and Weaknesses. *Workshop on*

Mathematical Modelling of Wave Phenomena with applications in the power industry, Linnaeus University, Växjö. http://lnu.se/mmwp.

[11] Kuvshinov, A.A; Abramov, G.N. (2003). Logico-Algebraic Modeling and Synthesis of Information and Power Structures of Multiphase Static Transducers in the Basis of Operations of Predicate Algebra of Choice. *Electromechanics, Proceedings of Universities*, No. 4, 29-36, (in Russian).

[12] Khrennikov, A.Yu. (2008). The problem of the electrodynamic stability of power transformers. *Industrial Energy*, № 9, 12-16, (in Russian).

[13] Khrennikov, A.Yu; Mazhurin, R.V. (2012). Powerful testing center in Russia. Technical and organizational factors. *News of Electrical Engineering*, № 3, (75), (in Russian).

[14] Khrennikov, A.Yu; Mazhurin, R.V. (2012). Electrodynamic tests to improve the reliability of transformer-reactor equipment (TRE), the existing testing base and the possibilities of TRE testing in the future. *ELECTRO*, № 5, (in Russian).

[15] Dementiev, Yu.A.; Smekalov, V.V., Shakaryan, Yu.G., Khrennikov, A.Yu., Kuvshinov, A.A. (2014). About establishment of the Federal Testing Center. Electrodynamic tests of power transformers for withstand to short-circuit currents. *News of Electrical Engineering*, № 1, (85), (in Russian).

[16] Kuvshinov, A.A.; Khrennikov, A.Yu. (2014). High-voltage thyristor valve for electrodynamic testing of power transformers. *ELEKTRO*, № 2, (in Russian).

[17] Khrennikov, A.Yu., Goldstein, V.G. Technical Diagnostics, Damage and Resources of Power and Measuring Transformers and Reactors. Moscow, *Energoatomizdat*, 2007, (in Russian).

[18] Konyukhov, A.V., Veselova, I.M., Martynenko, V.A., Khapugin, A.A., Grishanin, A.V. (2011). Domestic high-voltage power phototyristors with integrated self-protection from breakdown. *Electricity*, № 10, (in Russian).

[19] ABB Switzerland Ltd. (2002). *Semiconductors reserves the right to change specifications without notice.* Doc. No 5SYA1232-01.

[20] Wakeman, F., Li, G., Golland, A. (2005). New family of 4,5 kV Press-pack IGBTs. *PCIM'05*, Nuremberg, Germany.

[21] Abramovich, M.I., Babailov, V.M., Liber, V.E., Sakovich, A.A., Shper, V.L. (1992). Diodes and thyristors in converter installations. Moscow, *Energoatomizdat*, (in Russian).

[22] Voronin, P.A. (2001). *Power semiconductor switches: families, characteristics, application.* Moscow, Dodeka Publishing House-XX1 (in Russian).

[23] Kuvshinov, A.A., Khrennikov, A.Yu. (2014). Functional and switching features of a high-voltage high-current semiconductor key for electrodynamic testing of transformers. *Industrial Power*, № 9 (in Russian).

[24] Kuvshinov, A.A., Khrennikov, A.Yu. (2015). High-voltage high-current semiconductor key with direct light control for electrodynamic testing of power transformers. *Industrial Power*, № 2, (in Russian).

[25] Khrennikov, A.Yu., Schlegel, O.A. (2004). Control of the change in the inductive resistance of transformers to determine the damage in the windings. *Energetik*, № 2 (in Russian).

[26] Khrennikov, A.Yu. (2006). Information-measuring systems for monitoring and protecting windings of power transformers and reactors. *Scientific Bulletin of the National Technical University*, Novosibirsk, № 1 (22), ISSN 1814-1196 (in Russian).

[27] Patent of the Russian Federation №. 2041472 (1995). The device for testing the transformer short circuit currents. Khrennikov A. Yu., Lurie A. I., Schlegel O. A. *Bul.* № 22, 5.

[28] Patent of the Russian Federation № 2136099 (1999). A device for monitoring and protecting windings of transformers against deformation during short circuits. Khrennikov A. Yu., *Bul.* № 24, 10.

[29] Patent of Russian Federation, № 2063050 (1996). Control and protection device of transformer windings against deformation at short circuits in service. Khrennikov, A. Yu., Shlegel, O. A., Lurie, S. I., *Bulletin* № 18, 7.

[30] Patent of Russian Federation, № 96110718 (1998). Measuring device of the inductive transformer short-circuit impedance. Khrennikov, A. Yu., Shlegel, O. A., *Bulletin* № 12, 7 p.

[31] Carroll, E., Klaka, S., Linder, S. (1998). Thyristors IGCT. A New Approach to High-Power Electronics. *Elektrotekhnika*, №. 7.

[32] Backlund, B., Ebner, S. (2011). The wind power converter for tomorrow is already here. *EWEA*, Brussels, Belgium.

[33] Arnold, M., Wikström, T., Otani, Y., Stiasny, T. (2011). High-Temperature Operation of HPT+ IGCTs. *Presented at PCIM*, Nürnberg, Germany.

[34] Nistor, I., Wikström, T., Scheinert, M., Rahimo, M., Klaka, S. (2010). 10 kV IGCT rated at 3200 A, a new milestone in high power semiconductors. *PCIM*, Nürnberg, Germany.

[35] Vobeck, J., Siegrist, R., Arnold, M., Tugan, K. (2011). Large Area Fast Recovery Diode with Very High SOA Capability for IGCT Applications. *Presented at PCIM*, Nürnberg, Germany.

[36] Bates, D.J., Beccuti, A.G., Fotiou, I.A., Morari, M. (2008). An optimal control application in power electronics using numerical algebraic geometry. *American Control Conference* 2008, pp. 2221-2226.

[37] Kaipia, T., Peltoniemi, P., Lassila, J., Salonen, P., Partanen, J. (2008). Power electronics in Smartgrids-Impact on power system reliability. *IET-CIRED. CIRED Seminar SmartGrids for Distribution.*, p. 1-4.

[38] Cabal, C., Cid, A.P., Seguier, L., Estibals, B., Alonso, C. (2008). Improved photovoltaic conversion chain with interleaved method. *IEEE Power Electronics Specialists Conference (PESC 2008)*, 70-75.

[39] Sherstyuk, V. (2001). Transistor keys for power electronics devices - IGBT, MOSFET, or may be a bipolar transistor. *Electronic Components*, № 2, 59-65 & №. 3, 47-51 & №. 4, 62-66.

[40] IGBT Characteristics. *Application Note AN-983*. IR, 2012.

[41] Kiraly, L. (1993). 500V IGBTs replace MOSFETs at lower cost. *Design Tip* DT93-3. IR,

[42] Linder, S. (2007). Power semiconductor devices. *ABB Review*, № 4, 34-39 & № 1, 62-66.

[43] Wintrich, A., Ulrich, N., Reimann, T., Tursky, W. (2010). Application manual power semiconductors. *Semikron*, 464.

[44] Chou, W. (2011). Ultra-fast 1200V IGBTs reduce switching and conduction losses. *Power Electronics*, 31, 08.

[45] Silvestre, J. (2008). Half-bridge bidirectional DC-DC Converter for small Electric Vehicle. *International Symposium on Power Electronics, Electrical Drives, Automation and Motion (SPEEDAM 2008)*, 884-888.

[46] Li, J., Huang, X., Zhang, L., Zheng, T.Q. (2008). Research of DC-DC power supply for IGCT gate unit. *3rd IEEE Conference on Industrial Electronics and Applications (ICIEA 2008)*, 1766-1769.

[47] Ecklebe, A., Schulz, S., Lindemann, A. (2008). Investigation of two-step commutated resonant matrix converter supplying a contactless energy transmission system. *IEEE Power Electronics Specialists Conference (PESC 2008)*, 22-28.

[48] Tschirley, S., Bernet, S., Streit, P. (2008). Design and characteristics of reverse conducting 10-kV-IGCTs. *IEEE Power Electronics Specialists Conference (PESC 2008)*, 92-97.

[49] Moshtagh, J., Jannati, M., Baghaee, H.R., Nasr, E. (2008). A novel approach for online fault detection in HVDC converters. *12th International Middle-East Power System Conference (MEPCON 2008)*, 307-311.

Chapter 8

PROGRAMMABLE STATIC CONVERTERS FOR INTELLIGENT ELECTRICAL NETWORKS

A. A. Kuvshinov*, PhD and A. Yu. Khrennikov†, PhD

ABSTRACT

The authors consider the basic operations of implicative algebra for the choice of the logical-algebraic modeling of static converters, provided that the objective and predicate variables are identified with the parameters of the energy and information processes. The possibility of formalizing the problems of synthesizing static converters based on logical-algebraic models, which are reproduced instrumentally in the elemental basis of single-channel relators, is substantiated. Examples of logical-algebraic models of typical energy transformations are given. A generalized logical-algebraic model is proposed that allows one to formalize the procedure for synthesizing a programmable static converter. The examples of the program specification of the type of the electric energy conversion being carried out are given.

---

* Corresponding Author's E-mail: alekseikuvshinov@yandex.ru.
† Author's E-mail: ak2390@inbox.ru.

Keywords: programmable static converters, intelligent electrical networks, implicative algebra, logical-algebraic modeling, single-channel relators

8.1. Introduction

Intelligent electrical networks, characterized by the use of mechanisms for obtaining, storing and processing knowledge to implement the necessary modes of operation and control algorithms, impose stringent requirements on static converters (SC) of types and parameters of electrical energy. Among the traditional requirements for increasing the energy and mass-dimensions, improving the dynamic characteristics, the indicators of multifunctionality, invariance, robustness that predetermine the stability of SC to damage, malfunctions and failures, the ability to adapt to changes in the type and quality of the input electric power, to introduction of new components. Satisfying the above criteria, the SC acquires the properties inherent in intelligent systems, for the creation of which the principles of situational management and such information technologies as expert systems, artificial neural networks, fuzzy logic, genetic algorithms [1, 2] are used.

The design of programmable, intelligent static converters requires the solution of problems not only of system analysis, but also of circuit synthesis. The task of the analysis is to determine the output coordinates of SC with respect to a given structure and input variables, is completely formalized and is always solvable either by analytical methods [3] or by numerical computations [4]. The synthesis problem consists in determining the internal structure of SC on the given sets of input and output variables. In this formulation, the synthesis problem is not formalized and requires the use of heuristic techniques, which predetermines the ambiguity of the solutions obtained.

For a formalized description of the SC, the concept of a commutation function connecting the input u_{input}, i_{input} and output u_{output}, i_{output} coordinates

$$u_{\text{output}} = \Psi_{\text{SC}} \cdot u_{\text{input}},\ i_{\text{input}} = \Psi_{\text{SC}} \cdot i_{\text{output}} \quad (8.1)$$

and determining the type and quality of the transformation [3, 5].

However, the switching function does not reflect the structure of the power part and the control system of the SP and does not allow solving the problems of circuitry synthesis.

In [5], for a systematic description and formalization of synthesis problems, it is suggested to treat the SC as a functional system that realizes some set of transformation operators. However, to describe the transformation operators, an adequate mathematical apparatus is not proposed that would allow one to establish a one-to-one correspondence between the form of the transformation operator and the circuit structure for its implementation, and hence completely formalize the synthesis problems of SC.

At the same time, it is widely known that two-valued algebraic logic (Boolean algebra) serves as a theoretical basis for the formalized solution of problems of analysis and synthesis of digital systems. This gives grounds for assuming the expediency and the prospects for the extension of logical-algebraic methods to continuous-discrete systems, including SC.

In the present work, for the formalization of synthesis problems, a logical-algebraic method of system description of SC is proposed, based on the laws and properties of continual algebraic logics developed by L.I. Volgin and combined by a function - the axiom of weighted power means in a single metasystem, including classical two-valued algebraic logic [6].

The most implementative tasks of the system description and formalization of the SC synthesis procedures are the implicative choice algebra (ICA), the domains of which are two sets of analog variables, subject $Y = \{y_1, y_2, \ldots, y_n\}$ and predicate $X = \{x_1, x_2, \ldots, x_n\}$, and the basic operations are physically reproduced by single-channel relators. This allows, subject to identification of object variables with parameters of the energy process, and predicate variables with parameters of the information process, to form a system description of the SC in the form of logical-algebraic (LA) models written in the basis of the ICA operations, which can then be converted into electrical circuits by means of an elemental basis of relators [7]. As a result, the problem of synthesizing the SC is completely formalized.

8.2. Basic Operations of the Implicative Selection Algebra

Basic for ICA are binary operations

$$\vee(y_1, y_2) = y_1 \cdot I(x_1 - x_2) + y_2 \cdot I(x_2 - x_1), \tag{8.2}$$

$$\Lambda(y_1, y_2) = y_1 \cdot I(x_2 - x_1) + y_2 \cdot I(x_1 - x_2), \tag{8.3}$$

where y_1, y_2 - subject variables, which can serve as mathematical objects and physical parameters that satisfy the conditions of separation $y_i \cdot 1 = y_i$ and absorption $y_i \cdot 0 = 0$; x_1, x_2 - predicate variables (real numbers); $I(x)$ - the unit function (Heaviside operator), equal to 1 for $x > 0$ or 0 for $x < 0$.

Analysis (2), (3) allows us to conclude that the basic binary operations of ICA have implicativity properties, which is an analogue of the Boolean "EXCLUSIVE OR" function in the continual domain, and perform an alternative choice of one of the objective variables, either y_1, or y_2. It should also be noted that the procedure for identifying objective and predicate variables, for example $y_1 = x_1$, $y_2 = x_2$ is accompanied by binary values of the latter $x_i \in \{0,1\}$ by the degeneration of the basic ICA operations (8.2), (8.3) into logical operations of Boolean disjunction $x_1 \vee x_2$ and conjunction $x_1 \wedge x_2$, respectively.

Hardware implementation of basic ICA operations (8.2), (8.3), and when the above constraints and Boolean disjunctions and conjunctions are fulfilled, the analog logic element is implemented - a single-channel relator RL that contains a DA differential comparator that controls the two-operation keys of the closing S and breaking $\overline{S}$ types forming depending on the method of connection, the switching channel of the multiplexer (M-relator) or demultiplexer (D-relator) types, shown in Figure 8.1a, b, respectively [6]. The single-channel RL relator has a wide adaptive capability. In particular, increasing the load capacity of keys $S,\overline{S}$ and expanding, by introducing a digital element DD, algorithms for operating a switching channel in accordance with a family of reproducible ICA functions

$$W(y_1,y_2)=D_L\cdot[D_R\cdot\vee(y_1,y_2)+\overline{D}_R\cdot\wedge(y_1,y_2)], \tag{8.4}$$

allows to transform a single-channel RL relator into a power relator RLF (Figure 8.1c) [8]. As can be seen from (4), the digital signal D_L blocks $(D_L=0)$ or permits $(D_L=1)$ the execution of ICA operations (8.2), (8.3), and the digital signal $D_R=\{0,1\}$ ensures the interchange of ICA operations $(\vee\leftrightarrow\wedge)$.

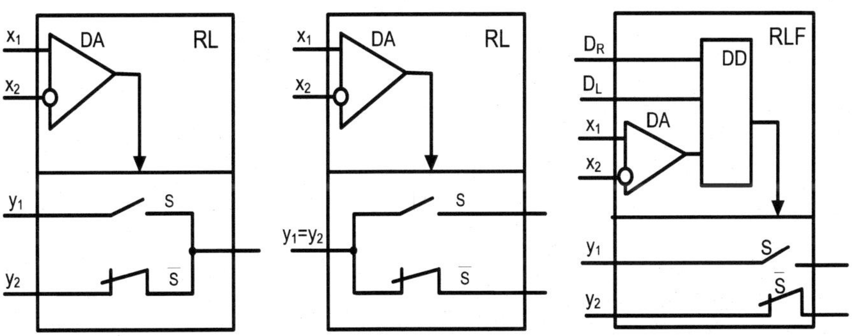

Figure 8.1. Structural diagrams of single-channel (a, b) and power (c) relators.

Analysis of (8.2), (8.3) and (8.4) allows us to note the principal possibility of simulating the simultaneous energy and information processes taking place in the SC in the basis of ICA operations, thanks to the use of two sets of variables that do not intersect in the general case - subject and predicate ones. It is sufficient to consider physical parameters of the energy process as objective variables, and control (information) signals as predicate variables.

8.3. LOGICAL-ALGEBRAIC MODELS OF ENERGY TRANSFORMATIONS

In the most typical for most single-phase SP case, only one source of input voltage is used and, accordingly, the following variants of identifying object variables

$$y_1 = e(t),\ \ y_2 = 0\ , \tag{8.5}$$

$$y_1 = e(t),\ \ y_2 = -e(t)\ , \tag{8.6}$$

where $e(t)$ is the input voltage of the SC, variable

$$e(t) = e(t+T) \tag{8.7}$$

With a period of variation of T or a constant

$$e(t) = \pm E. \tag{8.8}$$

When identifying subject and predicate variables

$$y_1 = e(t)\ ;\ y_2 = 0\ ;\ x_1 = y_1\ ;\ x_2 = y_2\ , \tag{8.9}$$

With input variable with the change of the sign of the voltage $e(t)$ basic binary ICA operations

$$V(y_1,y_2)=V(e,0)=e(t)\cdot I[e(t)-0]=\begin{cases}e(t) & npu\ e(t)>0\\ 0 & npu\ e(t)<0\end{cases}, \quad (8.10)$$

$$\Lambda(y_1,y_2)=\Lambda(e,0)=e(t)\cdot I[0-e(t)]=e(t)\cdot\{1-I[e(t)]\}=\begin{cases}0 & e(t)>0\\ e(t) & \text{for } e(t)<0\end{cases}, \quad (8.11)$$

determine the nature of the change in the output voltage of the SC and reproduce the LA models of uncontrolled single-phase half-wave rectification with positive and negative polarity of the rectified voltage, respectively. Analysis (8.10), (8.11) shows that the interchange of the basic binary operations $V \leftrightarrow \Lambda$ by, as follows from (8.2) and (8.3), the interchange of either subject variables $y_1 \leftrightarrow y_2$, or predicate variables $x_1 \leftrightarrow x_2$, is accompanied by a reversal of the polarity of the rectified voltage.

Based on the LA models (8.10), (8.11), the ICA functions can be obtained

$$V(e,0)-\Lambda(e,0)=e(t)\cdot I\ [e(t)]-e(t)\cdot\{1-I\cdot[e(t)]\}=e(t)\cdot Sign\ [e(t)]=|e(t)|, \quad (8.12)$$

$$\Lambda(e,0)-V(e,0)=e(t)\cdot\{1-I[e(t)]\}-e(t)\cdot I[e(t)]=-e(t)\cdot Sign[e(t)]=-|e(t)|, \quad (8.13)$$

which reproduce the LA models of unregulated single-phase full-wave rectification with positive and negative polarity of the rectified voltage, respectively. Here

$$Sign[e(t)]=\begin{cases}+1 & e(t)>0\\ -1 & \text{for } e(t)<0\end{cases}$$

there is a sign function.

A similar result can be obtained for modified conditions for identifying subject and predicate variables

$$y_1 = e(t);\ y_2 = -e(t);\ x_1 = y_1;\ x_2 = y_2 \tag{8.14}$$

In this case, the basic binary ICA operations (8.2), (8.3)

$$\begin{aligned} \mathrm{V}(y_1, y_2) &= \mathrm{V}(e,-e) = e(t)\cdot I[2\cdot e(t)] - e(t)\cdot I[-2\cdot e(t)] = \\ &= e(t)\cdot Sign[e(t)] = |e(t)|, \end{aligned} \tag{8.15}$$

$$\begin{aligned} \Lambda(y_1, y_2) &= \mathrm{V}(\mathrm{e},-e) = e(t)\cdot I[-2e(t)] - e(t)\cdot I[2\cdot e(t)] = \\ &= -e(t)\cdot Sign[e(t)] = -|e(t)|, \end{aligned} \tag{8.16}$$

also reproduce the LA models of unregulated single-phase full-wave rectification with positive and negative polarity of the rectified voltage, respectively.

Despite the same final result, the transformations (8.12), (8.13) and (8.15), (8.16) reflect different ways of realizing rectification processes. The rectified voltage under the conditions of identification (8.9) is formed by inversion according to the criterion of the sign of the input alternating voltage, which requires simultaneous execution of binary operations (8.2), (8.3) as follows from (8.12), (8.13). This method of conversion is carried out in bridge rectifiers. To form the rectified voltage under the identification conditions (8.14) it is sufficient, as follows from (8.15), (8.16), to perform one binary operation to carry out only an alternative choice by the criterion of the sign of one of the two antiphase variable voltages, since the inversion procedure was performed previously. This method of conversion is carried out in rectifiers with an average point.

8.4. LOGICAL-ALGEBRAIC MODELS OF PROGRAMMABLE ELECTRICITY TRANSFORMATIONS

Typical control algorithms of SC for different purposes are based on the procedure for comparing the reference signal $u_0(t)$ of the unfolding view and the signal of regulation $u_r(t)$, which generated by the automatic controller or specified by the operator [7]. Such a comparison will be carried out in the process of implementing the basic binary operations (8.2), (8.3), if these signals are identified with predicate variables, for example, $x_1 = u_0(t)$. The reference signal in the general case can be represented as a composition of two functions

$$u_0(t) = U_0(t) \cdot Sign[e(t)], \tag{8.17}$$

where $U_0(t)$ is the modulation function determining the shape of the reference signal;

$$Sign[e(t)] = \begin{cases} +1 & e(t) > 0 \\ -1 & \text{for } e(t) < 0 \end{cases}$$

- sign function that synchronizes the reference signal with the supply voltage.

When $U_0(t) = U_0 = const$ the reference signal has the form of a meander and provides only synchronization of the switching channels of the power RLF relator with the transitions of the input voltage through zero values. By choosing the law of variation $U_0(t) = \text{var}$, it is possible to give the reference signal practically any shape, which necessary for the formation of a given transfer characteristic of SC. Under the conditions (8.8), the reference signal, as follows from (8.9), must perform the function of the reference signal

$$u_0(t) = U_0(t) = U_0(t + T_{\text{con}}) = u_{\acute{K}}(t), \tag{8.18}$$

which determines the switching frequency of the switching channel of the power RLF relator and, correspondingly, the frequency of conversion of electrical energy $f_{\text{con}} = 1/T_{\text{con}}$.

If the objective (y_1, y_2) and predicate (x_1, x_2) variables satisfy the identification conditions $y_1 = e(t)$, $y_2 = 0$, $x_1 = u_0(t)$, $x_2 = u_r(t)$, then the basic binary operations (8.2), (8.3) reproduce the functions

$$\vee(y_1, y_2) = y_1 \cdot I(x_1 - x_2) = e(t) \cdot I\left[u_0(t) - u_r(t)\right], \tag{8.19}$$

$$\Lambda(y_1, y_2) = y_1 \cdot I(x_2 - x_1) = -e(t) \cdot I\left[u_0(t) - u_r(t)\right], \tag{8.20}$$

half-wave rectification with pulse-phase regulation of the output voltage, positive (8.19) or negative (8.20). The interchange of predicate variables ($x_1 \leftrightarrow x_2$) is accompanied, as can be seen from (8.19), (8.20), by reversing the polarity of the rectified voltage.

Under the constraints (8.14) for object variables (y_1, y_2) and preserving the identification conditions for predicate variables $x_1 = u_0(t)$, $x_2 = u_r(t)$, the basic binary operations (8.2), (8.3) reproduce functions

$$\vee(y_1, y_2) = y_1 \cdot [I(x_1 - x_2) - I(x_2 - x_1)] = e(t) \cdot Sign\left[u_0(t) - u_r(t)\right], \tag{8.21}$$

$$\Lambda(y_1, y_2) = y_1 \cdot [I(x_2 - x_1) - I(x_1 - x_2)] = -e(t) \cdot Sign\left[u_0(t) - u_r(t)\right], \tag{8.22}$$

full-wave rectification with pulse-phase regulation of the output voltage of positive or negative polarity, respectively. In the particular case where the regulation signal $u_r(t) = 0$, expressions (21) and (22) degenerate into

(8.15), (8.16), respectively, determining the limiting value of the rectified voltage.

The analysis (8.19), (8.20) under the conditions of identification of objective variables

$y_1 = |e(t)|$ or $y_1 = -|e(t)|$

and the identification of the predicate variable $x_1 = u_з(t)$ with the reference signal $u_з(t)$ of the unfolding form, whose frequency should exceed the frequency of the input voltage $e(t)$, shows that the basic operations (8.2), (8.3)

$$\mathrm{V}(y_1, y_2) = |e(t)| \cdot I[u_3(t) - u_r(t)] = \frac{1}{2} \cdot |e(t)| + \frac{1}{2}|e(t)| \cdot Sign[u_3(t) - u_r(t)] = \begin{cases} |e(t)| \\ 0 \end{cases}$$

$u_з > u_r$

$u_з < u_r$

for, (8.23)

$$\Lambda(y_1, y_2) = -|e(t)| \cdot I[u_з(t) - u_r(t)] = -\frac{1}{2} \cdot |e(t)| - \frac{1}{2} \cdot |e(t)| \cdot Sign \cdot [u_з(t) - u_r(t)] = \begin{cases} -|e(t)| \\ 0 \end{cases}$$

$u_з > u_r$

$u_з < u_r$

for , (8.24)

reproduce the functions of pulse-width regulation (PWR) of the rectified voltage of positive (8.23) or negative (8.24) polarity. The relations (8.23) and (8.24) are LA models of rectifiers with a PWR on the DC current side [8-15].

The analysis of (8.19), (8.20) when the condition (8.8) is satisfied for the subject variable y_1 and condition (8.18) for the predicate variable x_1

shows that the operations of the IAA disjunction (8.2) and the ICA-conjunction (8.3) [16-23].

$$V(y_1, y_2) = E \cdot I[u_з(t) - u_r(t)] = \frac{1}{2} \cdot E + \frac{1}{2} \cdot E \cdot Sign[u_з(t) - u_r(t)] = \begin{cases} E \\ 0 \end{cases}$$

$u_з > u_r$

for $u_з < u_r$, (8.25)

$$\Lambda(y_1, y_2) = -E \cdot I[u_з(t) - u_r(t)] = -\frac{1}{2} \cdot E - \frac{1}{2} \cdot E \cdot Sign[u_з(t) - u_r(t)] = \begin{cases} -E & u_з > u_r \\ 0 & \text{for } u_з < u_r \end{cases}, \quad (8.26)$$

reproduce the functions of pulse-width regulation of a constant voltage of positive (8.25) or negative (8.26) polarity.

Analysis (8.21) and (8.22) also allows us to note that each of the basic ICA operations (8.2), (8.3) under conditions (8.8), (8.18) reproduces the function of converting a DC voltage into an AC voltage with amplitude E and repetition period TR. The mutual replacement of predicate variables ($x_1 \leftrightarrow x_2$) and basic ICA operations ($V \leftrightarrow \Lambda$) is respectively accompanied by a change in the phase of the output alternating voltage to the opposite one.

Based on (8.21) and (8.22) with the use of (8.4), a generalized LA-model of the full-wave SC is formed as a function of the instantaneous values of the output voltage recorded in the predicate-logical form

$$u_{\text{output}}(t) = D_L \cdot \left\{ D_R \cdot \left[e(t) \cdot Sign\ (u_0(t) - u_r(t)) \right] + \overline{D}_R \cdot \left[(-1) \cdot e(t) \cdot Sign\ (u_0(t) - u_r(t)) \right] \right\}. \quad (8.27)$$

The generalized LA-model (8.27) shows the principal possibility of not only regulating the output voltage of the SC with the help of analog signals u_0, u_r, but also setting (programming) the required type of current conversion by interrupting, inversion, switching the level of the input voltage at individual time intervals in accordance with the formation algorithm of digital signals D_R, D_L.

The obtained LA models of various methods of unregulated and regulated rectification can be converted by means of an elemental basis of relators into electrical schemes of power and information structures of the SC, allowing to formalize the problems of circuit engineering synthesis [22, 23].

8.5. Examples of Programming of Energy Transformations

Analysis of the LA model (8.27) shows that the regulated rectification is provided by specifying digital signals $D_L = 1,\ D_R = 1$ or $D_R = 0$ depending on the required polarity of the rectified voltage, forming the in-phase with the supply voltage of the reference signal of the deployment form in accordance with (8.17) and fixing the operating point control signal u_r on the regulated characteristic, whose form is determined by the law of change of the modulation function $U_0(t)$.

The formation of a digital signal D_R in the form of a single function

$D_R = I[e(t + nT)]$ at n = 2, 3, 4

allows to transform the ICA function (27) into the LA model

$$W(y_1, y_2) = D_L \cdot |\, e(t)\,| \cdot Sign[e(t + nT)], \tag{8.28}$$

frequency conversion and generation of an output AC voltage, with a frequency “n” times lower than the frequency of the input voltage of the power supply.

An unregulated DC to AC conversion is provided by specifying digital signals $D_L = 1$, $D_R = 1$ or $D_R = 0$ by fixing a predicate variable $x_2 = u_r = 0$ and generating a reference signal in the form of a periodic function of arbitrary shape (8.18), for example, a meander that determines the switching frequency $f_{con} = 1/T_{con}$ of the switching channel. At the same time, an output alternating voltage of rectangular shape with amplitude E and a repetition period TП is formed.

For pulse-width regulation of the output voltage, it is necessary to provide the formation of a reference signal not only, for example,

$$u_0(t) = U_0 \cdot [1 - Sin(4\pi \cdot \frac{t}{T_\Pi})], \quad u_r = const \tag{8.29}$$

but also digital signals in the form of single functions

$$D_R = Sign[Sin\frac{2\pi}{T_\Pi} \cdot t], \quad D_L = I[u_0(t) - u_r]. \tag{8.30}$$

Then the ICA function (27) takes the form

$$W(y_1, y_2) = E \cdot Sign[Sin 2\pi \cdot \frac{t}{T_{con}}] \cdot I[u_0(t) - u_r] \tag{8.40}$$

As can be seen, only when $u_0(t) > u_r$ basic ICA operations are reproduced and an output voltage equal to $+E$ or $-E$ is generated. On the intervals $u_0(t) < u_r$, the switching channel is blocked, because $D_L = 0$, by interrupting the output voltage. The modulation of the digital signal D_R ensures a periodic interchange of the basic ICA operations (V $\leftrightarrow$ Λ), which is accompanied by an inversion of the polarity of the output voltage U_{output}.

The change in the control signal in the range $0 \le u_r \le 2 \cdot U_0$ is accompanied by a change in the half-wave length of the AC voltage from 0 up to $T_{con}/2$ and the output voltage from zero to the maximum value. The examples considered do not exhaust the possible combinations of the values of the master information signals u_o, u_r, D_R, D_L [22, 23].

CONCLUSION

A new approach to the system analysis of energy and information processes in single-phase static converters based on the logical-algebraic apparatus of implicative selection algebra is proposed. Its basic operations extend to the binary, discrete, continual range of values of variables and are instrumentally reproduced by universal circuit elements, single-channel relators.

The method of logical and algebraic modeling of single-phase static converters is developed in the basis of binary operations of the implicative choice algebra based on the procedures for identifying object and predicate variables with the parameters of information and energy processes and allowing to formulate on a single methodological basis logical-algebraic models reflecting the types of reproducible energy transformations, methods of program task of the current transformation, the structure of the electrical circuit.

ACKNOWLEDGMENTS

These studies are based on the fundamental scientific works of Ph.D., Professor, Honored Worker of Science and Technology of the Russian Federation Volgina L. I., in which the logical-algebraic apparatus of implicative choice algebra was developed, which served as the basis for logical-algebraic modeling and synthesis of programmable converters. The authors are grateful to the staff of the research laboratory of high-tech engineering at Ulyanovsk State Technical University for the useful discussions and continued support.

REFERENCES

[1] Uskov, A. A. Kuzmin, A. V. (2004) *Intelligent Control Technology. Artificial Neural Networks and Fuzzy Logic*. Moscow: Hot Line-Telecom, 143 p. (in Russian).

[2] Pupkov, K. A., Konkov, V. G. (2003). *Intellectual Systems*. Moscow: Publishing House MSTU. N. E. Bauman,348 p. (in Russian).

[3] Rudenko, V. S., Zhuikov, V. Y., Koroteev, I. E. (1980). *Calculation of Devices Converting Technology*. Kiev: Technique,. - 135 p. (in Russian).

[4] Dyakonov, V. P. (1998). *Handbook of MathCAD PLUS 7.0 PRO*. Moscow: SK Press, 102 p. (in Russian).

[5] Fogelberg, T., Girgis, R. S. (1994). ABB power transformers - a result of merging different technologies with prospects for significant future advancements. - *International Simposium "Electrotechnics – 2010,"* Moscow, vol.1.

[6] McNutt, W. J., Johnson, W. M., Nelson, R. A., Ayers, R. E. (1970). Power transformer short-circuit strength - requirements, design, and demonstration, *IEEE Trans. Power Appar. and Syst.*, 89, N 8.

[7] Malewski, R., Khrennikov, A. Yu. (1995). Monitoring of Winding Displacements in HV Transformers in Service. *CIGRE Working Group 33.03*. Italy, Padua, 4-9 Sept.

[8] Khrennikov, A. Yu., Goldshtein, V. G., Skladchikov, A. A. (2010). *The Analysis of a Condition of Overhead Lines of Power Transmission 6 – 500 Kv*, Power plants, № 5.

[9] Khrennikov, A. Yu., Goldshtein, V. G. (2007). The technical diagnostics, damages and resource of power and measurement transformers, reactors. *Energo-Atomizdat* (319 p.) (in Russian).

[10] Khrennikov, A. Yu. (2009) *Complex diagnostic modeling of technical parameters of power transformer- reactor electrical equipment condition*. PhD Thesis, The Samara University (in Russian).

[11] Khrennikov, A. Yu. (2004). Short-circuit performance of power transformers. Test experience, including fault diagnostic, *Reports of Advanced Research Workshop on Modern Transformers (ARWtr 2004)*, Vigo – Spain.

[12] Khrennikov A. Yu. New "intellectual networks" (Smart Grid) for detecting electrical equipment faults, defects and weaknesses. *Smart Grid and Renewable Energy*, February, 2012, Volume 3, Issue 1.

[13] Khrennikov, A. Yu. (2013). Smart Grid technologies for Detecting Electrical Equipment Faults, Defects and Weaknesses. *Workshop on Mathematical Modelling of Wave Phenomena with applications in the power industry*, Linnaeus University, Växjö. http://lnu.se/mmwp.

[14] Khrennikov, A. Yu., Shakaryan, Y. G., Dementyev, Y. A. (2013). Shortcurrent Testing Laboratories. Short-Circuit Performance of Power Transformers, Transformer Testing Experience. *International Journal of Automation and Control Engineering* (IJACE), Volume 2, Issue 3. (pp. 120-127). http://www.seipub.org/ijace/AllIssues.aspx.

[15] Khrennikov, A. Yu. (2014). Transformer Testing Experience by LVI/FRA Methods. Short-Circuit Performance of Power Transformers. *Current Advance in Energy Research* (CAIER). Volume 1, Issue 1 Jun. (pp. 8-14). www.vkingpub.com/VkUpload/20140613110141888l.pdf.

[16] Khrennikov, A. Yu., Mazhurin, R. V. (2014). The SF6 high-voltage electrical equipment: analysis of accident rate and service experience. *Current Advance in Energy Research* (CAIER) Volume 1, Number 2. (pp. 44-49). http://www.vkingpub.com/Journal/CAIER/245.html.

[17] Khrennikov, A. Yu., Mazhurin, R. V. (2014). High-voltage electrical equipment with the SF6 (sulfur hexafluoride) gas insulation: analysis of accident rate and service experience. *International Journal of Engineering Research & Management* (IJERM), Volume 1, Issue 5. (pp. 31-36). http://www.ijerm.com/ar_download.php?archive=Vol.1+Issue+5&&archive_id=9.

[18] Khrennikov, A. Yu., Mazhurin, R. V., Radin P. S. (2015). Diagnostic Methods for Detection of Electrical Equipment's Faults and Defects Including with Sulfur Hexafluoride Gas Insulation. *Energy and Power Engineering Science, EPES* 2015, 2(1): pp. 13-19. http://manuscript.sciknow.org/uploads/epes/pub/epes_14139061671.pdf.

[19] Khrennikov, A. Yu. Diagnostics of Electrical Equipment Faults and Power Overhead Transmission Line Condition by Monitoring Systems (Smart Grid): *Short-Circuit Testing of Power Transformers*, Nova publishers, New York, 2016, USA, p. 165 .https://www.novapublishers.com/catalog/product_info.php?products_id=56561.

[20] Khrennikov, A. Yu., Mazhurin, R. V., Radin, P .S., Skrydlov, E. I., Khrennikova, T. A. Estimation of the technical condition state of substation electrical equipment with the use of software and information tools, *Journal of Basic and Applied Research International,* 2015 - Volume 7 (Issue 3), – P. 201-210. http://www.ikpress.org/issue.php?iid=520&id=42.

[21] Khrennikov, A. Yu., Kuvshinov, A. A., Mazurov, V. A., Nesterchuk, A. K. Control of Power Transformer Windings during Electrodynamic Test Using High-Voltage Thyristor Valve for Regulation the Angle of the Initial Inclusion, *International Journal of Automation and Control Engineering,* Vol. 4, No. 2—October 2015, p. 58-66. www.seipub.org/ijace.

[22] Bulgakov, A A Generalized model of gate converters, *Electricity Journal*. No. 3, 1993, p.25–31. (in Russian).

[23] Kuvshinov, A. A, Abramov, G. N. Logico-Algebraic Modeling and Synthesis of Information and Power Structures of Multiphase Static Transducers in the Basis of Operations of Predicate Algebra of Choice. *Electromechanics, Proceedings of Universities,* No. 4, 2003, p. 29-36, (in Russian).

Chapter 9

CONTROLLED REACTOR TRANSFORMERS (CRT) WITH REDUCED HIGH CURRENT HARMONICS. THE CALCULATION OF PARAMETERS OF ACTIVE PARTS OF TRANSFORMERS BY DEFINING SPECIFIC ELECTROMAGNETIC LOADS

A. Yu. Khrennikov*, PhD and I. A. Shkuropat, PhD

ABSTRACT

The chapter presents Controlled Reactor Transformers with reduced high current harmonics. It is considered as a method of high current harmonics reduction in a thyristor-controlled reactor transformer. The high current harmonics reduction attains by combination of the LC- filters application and stepped smooth regulation. Below the high current harmonics reduction steps of the thyristor regulation are specified. The presence of the excess of reactive power in electrical lines causes negative

* Corresponding Author's E-mail: ak2390@inbox.ru.

consequences for the power system, such as operational breakdown of the electrical network. The presence of reactive power increases current consumption, which entails the increase in electrical losses, overloads. Controlled shunt reactor is intended to compensate the surplus capacitive power in transmission lines. In this case the inductive reactor power must be controlled and regulated simultaneously with the change of the excess capacitive power in the network.

One of the ways to control inductive reactance is a thyristor regulation in the reactor secondary winding. In this design of controlled shunt reactor transformer, changes in working inductance is carried out by shorting the secondary control winding through the thyristor key (TK). TK consists of two anti-parallel thyristors, which allow to eliminate even harmonics in the reactor. To implement smooth regulation of reactor power, the current is set by the opening angle of the thyristors. However, in this case, the incomplete opening of the thyristors leads to significant current distortion. As a rule, they are all odd harmonics starting from the 3rd one. The presence of distortion currents in the transformer windings, leads to additional heating of the the reactor windings.

Restriction of the level of high current harmonics in the reactor with smooth regulation of capacity is possible with the implementation of it in the form of multiple-winding transformer. The low-voltage winding is subdivided into several windings with its TK. The essence of the method of the high current harmonics reduction level is continuous, stepped regulation of the power of the controlled reactors, when all the previous control circuits are short-circuited, and the last one is smoothly adjustable by means of TK. In this case, high current harmonics in HVW, generated during TK operation in the last CW, are correlated not only with the current of the winding but also with the current of the first harmonic, caused by short-circuited CWs.

Keywords: controlled reactor transformer, thyristors, high current harmonics, stepped smooth regulation, LC – filters

9.1. INTRODUCTION

The surplus reactive power presence in electric transmission lines causes negative consequences for the power system, such as interruption of electric networks operation mode, deterioration of electric energy quality indices. The reactive power presence increases the current consumption. It leads to the increase of electric energy losses, overheating of feeding cables, transformers overloading and voltage levels are changed at the power consumers as well.

Controlled shunt reactors are designed for surplus capacitive power compensation in electric transmission lines and capacitor banks at electric energy consumers. In this case the induced power of reactor must be controlled and changed synchronously with change of surplus capacitive power in networks.

One of the ways of inductive resistance control in controlled reactors is thyristor regulation in secondary winding. In this design of controlled shunt reactor transformer (CSRT) change of operating inductance is performed by creating of counter magnetic flux in core by means of shorting of secondary control winding (CW) by means of thyristor key (TK). Herewith the reactor magnetic core is not saturated. TK consists of two anti-parallel connected thyristors that allow to eliminate even current harmonics in reactor.

To perform smooth regulation of reactor power current in CW is set as thyristor opening angle. However, in this case by not full opening thyristors occur significant current distortion. As a rule, there are odd harmonics starting from the 3rd one [1-3].

Current distortion presence, flowing in transformer winding, leads to increase of active resistance of transformer winding and additional heating. Overheating reduces transformer service life, and besides, the possibility to use transformer at full rated power is lost.

9.2. MEASURING MEANS OF ELECTRIC POWER QUALITY INDICES

Non-sinusoidal current flow can occur acceleration of insulation aging. It leads to service life decrease of electric equipment and electric transmission lines.

High harmonics create stray losses in transformers. These losses may lead to energy losses and cause transformers failure due to overheating.

Table 9.1. Measuring means of PQI and their characteristics

Measuring means of PQI / Characteristics of MM PQI	Proryv-KE	EM-3.2	Test-Electro	Fluke 435	ION 7650	MAVOWATT 70	SATEC PM175
Standard compliance	GOST 13109-97	GOST 13109-98	GOST 13109-99, IEC 61000-4-30	IEC 61000-4-30, EN50160	IEC 61000-4-30, EN50161	GOST 13109-97, IEC 61000-4-30, EN50162	GOST 13109-97, IEC 61000-4-30, EN50163
Measuring class	-	-	A (S)	A	A	A	A
Main interval of measuring time, s	NA	NA	NA	up to 0,5	0,2	0,2	0,2
United interval of measuring, s	3, 20, 60	3, 20, 60, 1800	NA	3, 600, 7200	3, 600, 7200	3, 600, 7200	3, 600, 7200
Measurement uncertainty of current time, ms/24h	NA	50	NA	20	20	20	20
Time synchronization	GPS (SNTP)	GPS	NA	NA	GPS	GPS	GPS
Marking of measurement results	NA	+	+	+	+	+	+

Table 9.2. Comparison of requirements of National Standards GOST R 51317.4.30-2008 and GOST 13109-97

PQI	GOST R 51317.4.30-2008				GOST 13109-97	
	Class A		Class S			
	Measurement uncertainty	**Measurement range**	**Measurement uncertainty**	**Measurement range**	**Limit of permissible measurement error**	**Measurement range**
Value and deviation (steady voltage deviation)	± 0,1%	10 - 150% Udin, where U_{din}	± 0,5%	20 - 120% U_{din}	± 0,2%	± 20%
Value and deviation of frequency	± 0,01 Hz	42,5 - 57,5/51 - 69 Hz	± 0,05 Hz	42,5 - 57,5/51 - 69 Hz	± 0,03 Hz	45 - 55 Hz
Factor of nth harmonic component	±5 % U_n at $U_n \geq 1$ % U_{rated}.; ±0,05 % U_n where $U_n < 1$ % U_{rated}	From 10 to 200% value level electromagnetical environment compatibility class 3 as per GOST R 51317.2.4	± 10% U_n where $U_n \geq$ 3% U_{rated}; ± 0,3% U_n where $U_n <$ 3% U_{rated}	From 10 to 200% value level electromagnetical environment compatibility class 3 as per GOST R 51317.2.4	± 0,03% where $K_u(n) < 1$, ± 3% where $K_u(n) \geq 1$,	0,05 - 30%
Voltage unbalance factor of negative sequence	± 0,15%	1 - 5% U_1	± 0,3%	1 - 5% U_1	± 0,2%	0 - 15%

Quantitative assessment of high harmonics component presence may be expressed in such electric energy quality indice, as the factor of nth harmonic component.

It is possible to measure the factor of nth harmonic component by means of specialized measuring instruments – Measuring means (MM) of power quality indices (PQI). Currently a great number of russian and foreign MM PQI are created. For example, such companies as OJS NPP "Proryv", Mars-Energo, NPP "TestElectro", Fluke, Schneider Electric, Meratest, SATEC, Siemens create MM PQI.

There are international standards IEC 61000-4-7:2002 and IEC 61000-4-30:2002 and National Standards GOST 51317.4.30-2008 and GOST 51317.4.7-2008 are developed on their base accordingly. In new measurement standards PQI are characterized as class A and S. The range of MM PQI was developed in accordance with GOST 13109-97, but the characteristics of MM PQI do not comply with the new standards. MM PQI and its characteristics are given in Table 9.1. In Table 9.2 are presented requirements comparison to measurement accuracy as per National Standards GOST 13109-97, GOST 51317.4.30-2008, GOST 51317.4.7-2008 for some PQI [4-6].

In Table 9.2. the following designations are used:

U_{din} – input voltage, specified by the voltage in electric supply system and measuring converter conversion factor.

U_n – rated voltage used to designate or identify electric supply system.

$K_u(n)$ – factor of nth voltage harmonic component.

U_l- negative sequence voltage.

9.3. Limitation of the Level of the High Current Harmonics. The Electrical Diagram of a Controlled Shunt Reactor of a Transformer Type

One of the effective ways of high current harmonics (HCH) restraint is of course application of LC-filters tuned to reduce certain harmonics. The most significant harmonics are 5th, 7th, 11th and 13th. However, this solution implies at least the use of four additional LC- filters to the reactor equipment.

Limitation of HCH level in reactor with smooth power regulation is quite possible by realizing it as the multi-winding transformer, which power high-voltage winding (HVW) is connected to alternative current network, and low-voltage winding (LVW) is subdivided into several CW with their TK. Each CW forms together with TK control circuit (CC) (Figure 9.1). Relative SC impedance between HVW and CW must be at least one.

HVW is primary and LVW is CW (Figure 9.1).

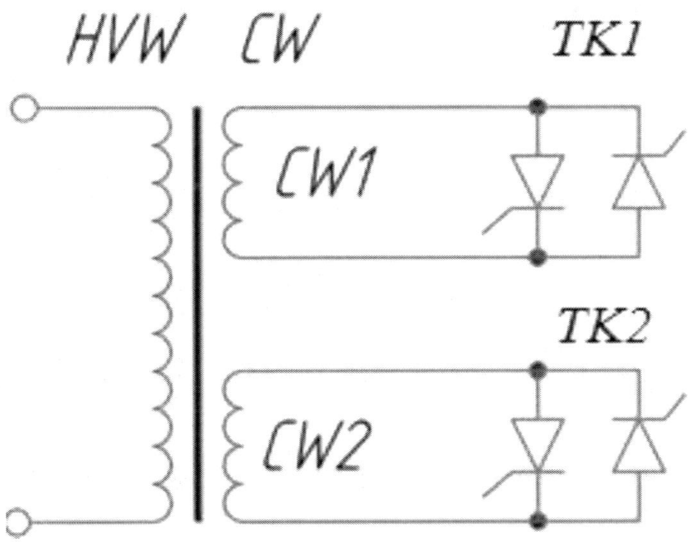

Figure 9.1. Single-line scheme of CRT.

The essence of the method of HCH level reduction consists in continuous step-by-step regulation of the power of CRT, when all previous CCs are short-circuited, and the last one is smoothly adjustable by means of TK. In this case, high current harmonics in HVW, generated during operation of TK in the last CW, are correlated not only with this winding current, but also with the current of the first harmonic, caused by short-circuited CW [1-7].

9.4. Operation of a Controlled Shunt Reactor of a Transformer Type

Let us consider in detail the process of smooth power regulation of CSRT.

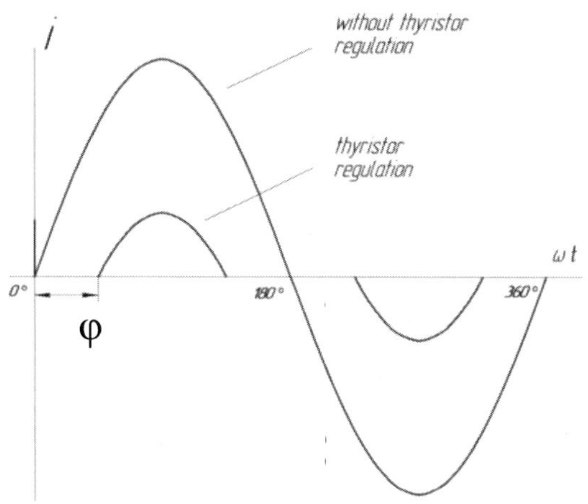

Figure 9.2. Current waveforms in CSRT windings by thyristor regulation on the first stage.

On the first, low regulation stage, current regulation is made by means of TK from no-load to full power, TK of this stage shall be short-circuited. In absolute value HCH content here is not high.

Current in first CC, regulated by means of TK, during period is changed according to the law depending on the opening angle φ (Figure 9.2):

$$i_1 = \begin{cases} I_{1st.m}\,[\sin(\omega t) - \sin(\varphi)], & \varphi < \omega t < \pi - \varphi\ ; \\ 0, & 0 < \omega t < \varphi \text{ и } \pi - \varphi < \omega t < \pi + \varphi\ ; \\ I_{1st.m}\,[\sin(\omega t) + \sin(\varphi)], & \pi + \varphi < \omega t < 2\pi - \varphi\ ; \end{cases} \tag{9.1}$$

where $I_{1st.m}$ — current amplitude on the first, thyristor-regulated stage, ω — circular power frequency; $\omega = 2\pi f$, f – power frequency, t — time, φ — thyristor opening angle.

Current is non-sinusoidal here. There are odd current harmonics. There are no even harmonics due to thyristor opening symmetry.

Having applied harmonic Fouriere analysis [3] to expression (9.1), we receive the expression of maximum value of fundamental current harmonics:

$$I_{m1} = \frac{I_{1st.m}}{\pi}(\pi - 2\varphi - \sin(2\varphi)). \tag{9.2}$$

Effective current of the first stage:

$$I_1 = \sqrt{\int_0^{2\pi} (i_1^2)\, d(\omega t)} =$$

$$= \frac{I_{1ст.m}}{\pi\sqrt{2}}\sqrt{\pi - 2\varphi - 3\sin(2\varphi) + (2\pi - 4\varphi)\sin^2(\varphi)}\,. \tag{9.3}$$

Total HHF shall be equal to:

$$k_h = \frac{\sqrt{I^2 - 0.5 I_{m1}^2}}{I_1} \tag{9.4}$$

HCH percentage content of the first stage:

$$k_г\% = k_{гх}100\%, \tag{9.5}$$

Dependence of HCH content on thyristor opening angle is calculated by formulae (2) – (5) and presented in Table 9.3.

HCH content depending upon thyristor opening angle and power gain is given In Table 9.4.

With a fully open TK with $\varphi = \pi/2$ in HVW flows sinusoidal current:

$$i_1 = I_{1st.m} * \sin(\omega t), \tag{9.6}$$

This current is "auxiliary" current for not activated second stage. It serves to decrease the relative HCH level, generated during the operation of the second CC.

Reactor operates in lower power range, from 0 to 5 % of rated power. Though the relative HCH content here is high, in their absolute HCH value content on the first stage does not exceed 5 % of rated current that is quite acceptable.

From this moment the reactor starts to operate on the second regulation stage. Its power is smoothly regulated by changing a thyristor opening angle and reach the maximum value when TK is fully open.

Current in second, regulated, CC during the period is changed according to the law (Figure 9.2):

$$i_2 = \begin{cases} I_{2st.m}[\sin(\omega t) - \sin(\varphi)], & \varphi < \omega t < \pi - \varphi ; \\ 0, & 0 < \omega t < \varphi \text{ and } \pi - \varphi < \omega t < \pi + \varphi ; \\ I_{2st.m}[\sin(\omega t) + \sin(\varphi)], & \pi + \varphi < \omega t < 2\pi - \varphi \end{cases} \tag{9.7}$$

where $I_{2st.m}$ – amplitude of current on the second, thyristor-regulated stage.

Thyristor opening angle is considered here from current transition zero-crossing moment in positive half-period. As a result of applying two anti-parallel thyristors connected in TK, the current waveforms i2 in positive and negative half-periods are symmetrical.

Operating current of reactor in HVW when operating on the second regulation stage consists of sinusoidal current in the first CC and non-sinusoidal current in the second CC (Figure 9.3):

$$i = I_{1st.m}\sin(\omega t) + i_2(\omega t), \tag{9.8}$$

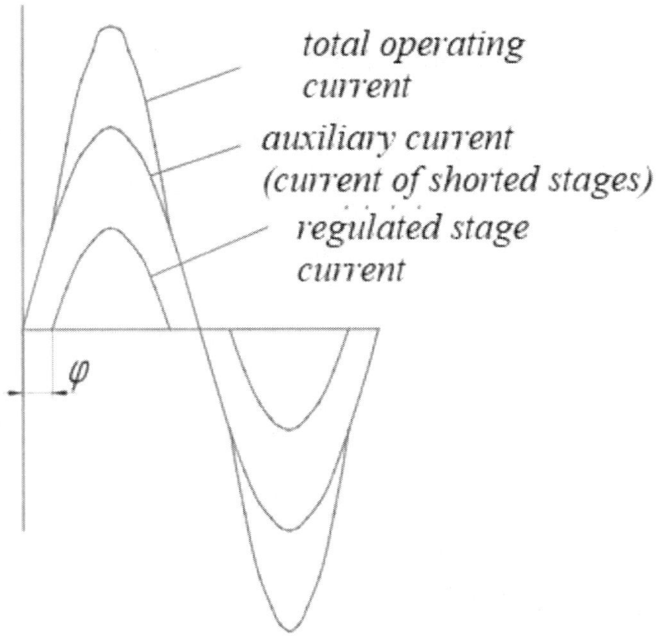

Figure 9.3. Current waveforms in CSRT windings by thyristor regulation of the last stage.

In CSRT with power splitting for stages to decrease HCH content is used the "auxiliary" current. Relative HCH value is determinated as power (current) ratio of the regulated stage to total power ("auxiliary current") of shorted stages - power gain. In this case we can see power gain for two stages:

$$K_{st} = I_{2st.m}/I_{1st.m}, \tag{9.9}$$

By fully closed second, regulated stage ($\varphi = \pi/2$) operating current in HVW is already sinusoidal and in its turn, "auxiliary" current for operation on the third regulation stage, when the first and the second CCs are short-circuited, and the third stage is smoothly regulated by means of TK.

Expanding (9.1) in Fourier series[4] and executing with reference to (9.3) necessary mathematical transformations, we receive the principal expression for the factor of n-th harmonic of operating current in HVW:

$$K_{hn} = \frac{K_{st}}{\pi n} \frac{\frac{\sin((n-1)\varphi)}{n-1} + \frac{\sin((n+1)\varphi)}{n+1}}{1 + \frac{K_{st}}{\pi}(\pi - 2\varphi - \sin(2\varphi))} \quad (9.10)$$

where n –harmonics number of operating current in HVW.

Table 9.3. Content of HCH depending on thyristor opening angle and power gain

Regulation angle φ°	High harmonic factor				
	One stage	Two stages K_{st} = 1.0		Two stages K_{st} = 2.2	
		without filters	LC-filter 5th harmonic	without filters	LC-filter 5th harmonic
0	0.00	0.00	0.00	0.00	0.00
10	11.07	4.72	2.69	9.26	5.27
20	23.01	5.34	1.94	9.43	3.43
30	36.28	2.99	1.15	4.81	1.84
40	51.62	2.33	2.20	3.44	3.26
50	70.35	3.02	0.75	4.13	1.03
60	95.20	2.99	1.15	3.79	1.45
70	133.0	1.61	0.95	1.91	1.13
80	210.6	0.35	0.28	0.39	0.31
90		0.00	0.00	0.00	0.00

Table 9.4. Distribution of CRT power with filter of the 5th harmonics by stages

Regulation stage	Without filter of the 5th harmonics		With filter of the 5th harmonics	
	Stage power, %	Total power of working stages, %	Stage power, %	Total power of working stages, %
1	5.0	5.0	6.0	6.0
2	6.0	11.0	20.0	26.0
3	13.0	24.0	74.0	100.0
4	26.0	50.0		
5	50.0	100.0		

Thus, according to (9.9 to 9.10) nth harmonic factor of operating current depends only on power gain of the stages and TK opening angle of the regulated stage.

In a three-phase reactor with thyristor regulation there are no even harmonics of operating current and multiple of three harmonics circulate in compensating windings connected in delta [2-3].

In operating current of CRT there are odd harmonics 5th, 7th, 11th and 13th with corresponding relative content k_{h5}, k_{h7}, k_{h11}, k_{h13}, calculated by(9.4).

Total HHF in this case is:

$$k_{h\Sigma} = \sqrt{(k_{h5}^2 + k_{h7}^2 + k_{h11}^2 + k_{h13}^2)}, \quad (9.11)$$

Calculations of dependance of total HHF on thyristor opening angle φ and power gain K_{st} are given in Table 9.3.

With power gain of stages $K_{st} = 1.0$, that is, when the power of the regulated stage is equal to the power of the previous shorted stages, total HHF does not exceed 5%. In this case reactor power may be subdivided into 5 stages according to Table 9.4.

9.5. Controlled Reactor of a Transformer Type with a Filter-Compensating Circuit

To decrease the number of regulation stages, it is advisable to apply LC-filter together with CRT for the most significant 5th harmonic. It allows to use not more than three regulation stages. Beyond that point at network frequency LC-filter serves as as a capacitor, compensating reactive power in the network.

CRT diagram with a LC-filter on HVW side is shown in Figure 9.4.

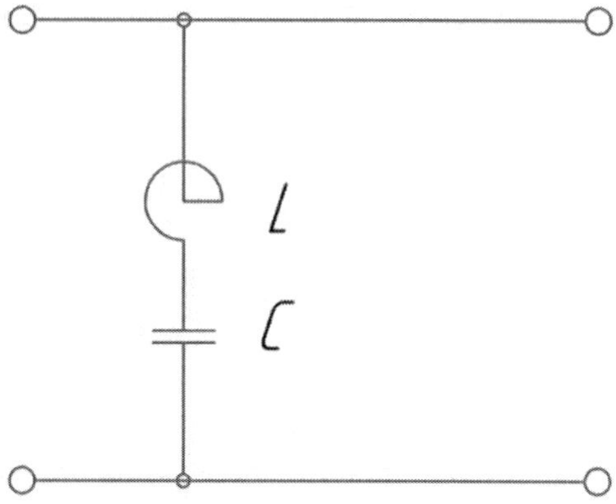

Figure 9.4. CRT with a LC- filter on HVW side.

By applying LC-filter of the 5th harmonics total HCH content decreases and there is the possibility to increase stage power gain. Thereby, the number of regulation stages will decrease.

Total HHF in this case is:

$$k_{h\Sigma} = \sqrt{(k_{h7}^2 + k_{h11}^2 + k_{h13}^2)}, \qquad (9.12)$$

The total HHF by power gain of the stages $K_{st} = 2.2$ and using LC-filter of the 5th harmonics is calculated by formulae (8.4) and (86). Calculation results are given in Table 9.3.

Let us calculate total HHF error. According to formula (12) 7th, 11th and 13th harmonics components have errors. Measuring these factors by means of MM PQI of class A as per GOST R 51317.4.30-2008, the ratio error of nth harmonics components does not exceed 5% where $k_{hn} > 1\%$ of U_r. The formula for determination of total error is given below [5-6]:

$$\delta k_{h\Sigma} = (k_{h7}^2 \cdot \delta k_{h7} + k_{h11}^2 \cdot \delta k_{h11} + k_{h13}^2 \cdot \delta k_{h13})/k_{h\Sigma}^2, \qquad (9.13)$$

Maximum value of total HHF does not exceed 5%, but the reactor power in this case can be subdivided already only into 3 stages: 6%, 20% and 74%.

It should be noted that TK in this case has different voltage classes. To reduce the price of the reactor TK one should decrese the voltage and accordinly increase the current in CW.

9.6. The Design of the Active Part of the Transformer

At this time there are a lot of diverse power transformer designs in transformer production industry. For example, there are transformers with noncircular shape of core leg and controlled reactors of transformer type with split windings.

It requires the creation of methodology for calculation of the transformer design. In this regard, the mathematical formalization of the relationships between the transformer design parameters and their electromagnetic loads on the active materials does not lose its relevance [1, 2].

The transformer calculation includes as a rule approximate technical commercial calculations to assess the selection of the optimal variant and accurate calculations that serve as the basis for detailed design of the transformer.

As it was already shown in [1], the fundamental principle for the design of the transformer is the power equation, uniquely connecting the basic dimensions with the electromagnetic parameters of the transformer active part.

Electrical equipment and transformers are inherently significant concentrators of electromagnetic energy. This energy is concentrated mostly in the air gaps of the electric motors and generators, interwinding leakage channel of the transformer. Specific density of this energy, expressed through the universal Arnold constant, is determined by specific electromagnetic loads: magnetic induction amplitude in core leg and current density in the windings [1]. Arnold constant expresses the relationship between the rated power and the specific electromagnetic loads. It is the initial base for electromagnetic calculations of electric equipment.

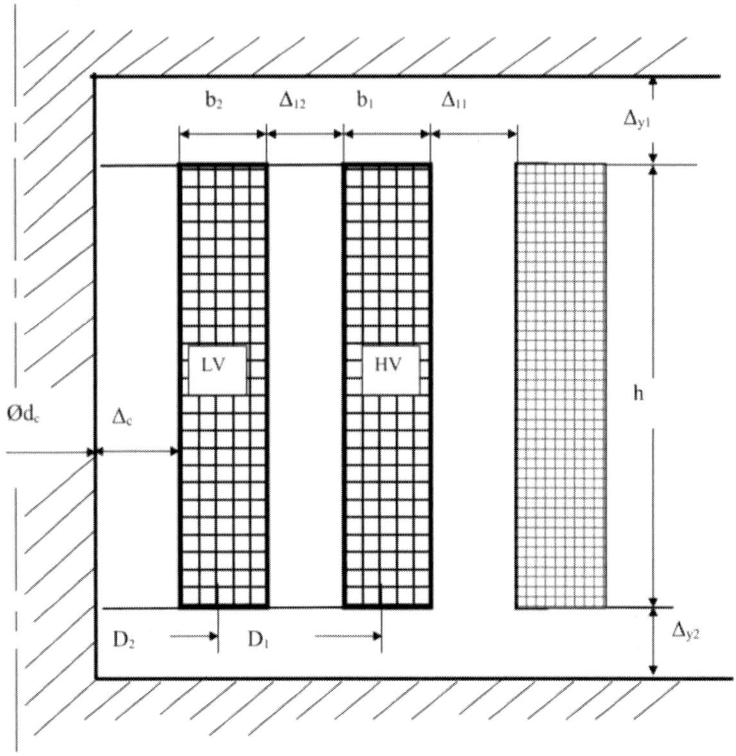

Figure 9.5. Windings arrangement in magnetic core window.

The meaning of the Arnold constant is that it accelerates the search for the basic dimensions of electrical equipment. For this purpose, based on the expression of the electromagnetic (design) power a conditional volume of the active part is determined associated with the basic dimensions, specific electromagnetic loads and power frequency.

The connection between the power transformer and specific electromagnetic loads on the active part was shown in general terms in [1].

Specific electromagnetic loads are the magnetic induction amplitude in the core leg in no-load mode, rated current density in windings.

The basic dimensions and parameters of the transformer active part are related with the equation of the electromagnetic power (equation of the electromagnetic usage of the active volume) with rated data and specific electromagnetic loads on the active materials.

The basic dimensions of the active part of a two-winding transformer are shown in Figure 9.5. Accumulated experience of electromagnetic, electricoinsulating and thermal calculations of the transformers of certain capacities and voltage classes help to choose the required values. Insulation-design dimensions specified in the calculation based on the experience in the design of transformers with a certain isolation voltage class [3] are the following (Figure 9.5).

Δ_{y1} — insulation distance between the top yoke and the ends of the windings with wire, Δ_{y2} — insulation distance between the bottom yoke and the ends of the windings with wire, Δ_c —insulation distance between the magnetic core and LV winding; Δ_{12} — insulation distance (the width of the leakage channel) between HV and LV windings, Δ_{11} — insulation distance between the external HV windings of adjacent phases.

Table 9.5. The insulation distances in transformers, mm

Insulation distances, mm	**Designation**	**Insulation voltage class**				
		6 kV	**10 kV**	**15 kV**	**20 kV**	**35 kV**
between the top yoke and windings ends with wire	Δ_{y1}	20	20	20	20	20
between the bottom yoke and windings ends with wire	Δ_{y2}	20	20	20	20	20
between core leg and LV winding	Δ_c	5	5	5	5	5
between HV and LV windings	Δ_{12}	8.5	8.5	8.5	8.5	8.5
between HV windings of different phases	Δ_{11}	10	10	15	20	30

Recommended values of insulation distances depending on the insulation voltage class are given in Table 9.5.

The rated values of the transformer are connected directly with the geometry of the active part, accepted specific electromagnetic loads, and other parameters.

9.7. Calculation of the Design of the Active Part of a Two-Winding Transformer

Let us derive the expression for the Arnold constant with regard to the transformers. Let us consider the simplest design of the active part of a two-winding transformer (Figure 9.5).

Magnetic core is flat-laminated. The cross section of the leg is round stepped. The design of the winding is the following. The winding of the low -voltage (LV) is internal, the winding of the high-voltage (HV) is external. The windings have equal height. The arrangement on the core leg is concentric.

Transformer power per phase is expressed as follows:

$$S_{r.p} = S_r/m_p = U_{r1p}I_{r1p} = U_{r2p}I_{r2p}, \tag{9.14}$$

where S_r – full rated power of the transformer, m_p – phase number, U_{r1p} and U_{r2p} – rated phase voltages of HV and LV windings accordingly, I_{r1p} and I_{r2p} – rated phase currents of HV and LV windings accordingly.

Rated phase voltages of HV and LV windings, rated phase currents of HV and LV windings are directly related with the transformer active part geometry.

Rated phase voltages of HV winding:

$$U_{r1p} = (\omega/\sqrt{2})w_1B_mS_c, \tag{9.15}$$

where ω — circular power frequency; ω = 2πf, f – power frequency w_1 — HV winding turn number; B_m — the magnetic induction amplitude in the core; S_c — active cross-section area of round core leg.

Rated phase voltages of LV winding:

$$U_{r2p} = (\omega/\sqrt{2})w_2B_mS_c, \tag{9.16}$$

where w_2 — turn number of LV windings.

Active cross-section area of round core leg:

$$S_c = k_c k_{geom}(\pi/4)d_c^2, \quad (9.17)$$

where d_c — circumscribed circle diameter of the core leg d_c, k_c — stacking factor of the core leg, approximately $k_c = 0.965$, k_{geom} — geometrical stacking factor of circular cross-section of the core leg, approximately $k_{geom} = 0.925$.

If in the transformer no-load current is negligibly small, then the magnetic moving forces (MMF) of HV and LV windings accordingly are arithmetically equal between each other:

$$w_1 I_{r1p} = w_2 I_{r2p}. \quad (9.18)$$

MMF of windings are related with windings geometry in the following way.

MMF of HV winding:

$$w_1 I_{r1p} = w_1 S_{t1} J_{r1}, \quad (9.19)$$

where S_{t1} — active cross-section area of HV winding turn, J_{r1} — rated current density of HV winding.

Active cross-section area of HV winding turn:

$$S_{t1} = I_{r1p}/J_{r1}, \quad (9.20)$$

MMF of LV winding:

$$w_1 I_{r1p} = w_2 S_{t2} J_{r2}, \quad (9.21)$$

where S_{t2} — active cross-section area of LV winding turn, J_{r2} — rated current density of LV winding.

Active cross-section area of LV winding turn:

$$S_{t2} = I_{r2p}/J_{r2}. \quad (9.22)$$

Active cross-section area of HV winding:

$$S_{w1} = k_{w1} w_1 S_{t1}, \tag{9.23}$$

where k_{w1} —filling factor of cross section of HV winding.

Active cross-section area of LV winding:

$$S_{w2} = k_{w2} w_2 S_{t2}, \tag{9.24}$$

where k_{w2} —filling factor of cross section of LV winding.

Taking into account expressions (9.15) - (9.22), transformer phase power is presented in the following way:

$$S_{r.p} = U_{r1p} I_{r1p} = \\ = (\omega/\sqrt{2}) B_m S_c w_1 I_{r1p}, \tag{9.25}$$

Determination of SC parameters.

SC impedance between HV and LV windings consists of inductive and active components:

$$Z_{sc} = \sqrt{X_{sc}^2 + R_{sc}^2}, \tag{9.26}$$

where R_{sc} — phase active SC impedance between HV and LV windings; X_{sc} — phase
inductive SC impedance between HV and LV windings:

$$R_{sc} = k_{add} \rho_{75} (\pi D_2 w_2 / S_{t2} + \pi D_1 w_1 / S_{t1}) = \\ = k_{add} \rho_{75} \pi \sqrt{2} (D_2 U_{r2p} J_{r2} / I_{r2p} + D_1 U_{r1p} J_{r1} / I_{r1p}) / (\omega B_m S_c), \tag{9.27}$$

$$X_{sc} = \omega (k_\rho \mu_0 w_1^2 S_{\sigma sc}) / h, \tag{9.28}$$

where k_{add} — factor of additional losses in transformer windings (SC losses ratio to main losses in windings), approximately k_{add}= 1.15, h – height of HV and LV windings.

Assuming that HV and LV windings have equal height (Figure 9.5) we can find their radial sizes. Rated phase voltages of HV and LV windings, rated phase currents of HV and LV windings are directly related with the transformer active part geometry.

Radial size of HV winding, with regard to (9.21), (9.24), (9.26):

$$b_1 = S_{w1}/h = k_{w1}w_1I_{r1p}/(J_{r1}h) = \sqrt{2}k_{w1}S_{r.p}/(\omega B_m S_c J_{r1}h). \quad (9.29)$$

Radial size of LV winding, with regard to (9.23), (9.25), (9.26):

$$b_2=S_{w2}/h = k_{w2}w_1I_{r1p}/(J_{r2}h) = \sqrt{2}k_{w2}S_{r.p}/(\omega B_m S_c J_{r2}h) \quad (9.30)$$

Rated current is calculated from (9.14) by means of SC parameters, such as SC impedance and rated phase voltage [8]:

$$I_{r1p} = S_{r.p}/U_{r1p} = (U_{r1p}/Z_{sc})(u_{sc}/100) \quad (9.31)$$

where u_{sc} — SC impedance (%) between HV and LV windings.

Phase inductive SC impedance between HV and LV windings:

$$X_{sc} = (\omega k_\rho \mu_0 w_1^2 S_{\sigma sc})/h, \quad (9.32)$$

where $S_{\sigma sc}$ — equivalent cross-section area of magnetic flux leakage channel between HV and LV windings, k_ρ — Rogowski factor (reduction factor of real magnetic field to ideal), μ_0 = 4π*10-7, H/m — universal magnetic constant.

Equivalent cross-section area of magnetic flux leakage channel between HV and LV windings:

$$S_{\sigma sc} = \pi D_{12}(b2/3 + \Delta_{12} + b1/3), \quad (9.33)$$

where D_{12} is mean diameter of the channel between HV and LV windings (Figure 9.5).

Rogowski factor:

$$k_\rho = 1 - (b_2 + \Delta_{12} + b_1)/(\pi^*h). \quad (9.34)$$

For the power transformer it is possible to neglect the active resistance and assume:

$Z_{sc} = X_{sc}$.

Then transformer phase power is presented in the following way:

$S_{r.p} = (u_{sc}/100)U_{r1p}{}^2/X_{sc}$. (9.35)

This follows the windings mean diameters.
Mean diameter of LV winding:

$D_2 = d_c + 2\Delta_c + b_2$. (9.36)

Mean diameter of the HV – LV channel:

$D_{12} = D_2 + b_2 + \Delta_{12}$. (9.37)

Mean diameter of HV windings:

$D_1 = D_2 + b_2 + 2\Delta_{12} + b_1$. (9.38)

9.8. Power Equation of a Transformer

Power transformer equation structural diagram [7-8], expressing the relation between basic dimensions and transformer electromagnetic parameters is the following:

$$\begin{cases} S_{н.ф} = \frac{\omega B_m^2}{2\mu_0} \frac{S_{\sigma к} h}{k_p} \frac{S_c^2}{S_{\sigma к}^2} \frac{u_к\%}{100\%}; \\ k_p = 1 - \frac{b_1+\Delta_{12}+b_2}{\pi h}; \\ S_{\sigma к} = \pi D_{12}\left(\frac{b_1}{3} + \Delta_{12} + \frac{b_2}{3}\right); \\ b_1 = \frac{\sqrt{2}\, S_{н.ф}}{\omega B_m S_c J_1 h k_{об1}}; \\ b_2 = \frac{\sqrt{2}\, S_{н.ф}}{\omega B_m S_c J_2 h k_{об2}}; \\ D_2 = d_c + 2\Delta_c + b_2; \\ D_1 = D_2 + b_2 + 2\Delta_{12} + b_1; \\ D_{12} = D_2 + b_2 + \Delta_{12}. \end{cases} \quad (9.39)$$

Transformer power equation, presented as a system (39), uniquely connects the active part geometry (windings height and diameters, active cross-section area of core leg) and accepted specific electromagnetic loads.

First equation in the system (9.39) is the principal. It expresses accumulated inductive power in interwindings space. Let us reveal the physical meaning of factors in this equation.

$$\frac{\omega B_m^2}{2\mu_0}$$

- magnetic field energy density, accumulated in magnetic core legs.

$$\frac{S_{\sigma к} h}{k_p}$$

- magnetic field volume in interwindings space.

$$\frac{\omega B_m^2}{2\mu_0} \frac{S_c^2}{S_{\sigma к}^2}$$

- magnetic field energy density in interwindings space.

Remaining equations in the system (39) reveal first expression.

Based on electromagnetic loads numerical values and phase power values are calculated the active part geometry. The Arnold constant for the transformer determinates the calculation method of the two-winding transformer.

The transformer calculation sequence is the following:

determination of active part general design;

determination of main electrical values: line and phase currents and voltages of HV and LV windings,

Basic dimensions calculation,

selection of magnetic core design;

selection of laminated steel type,

magnetic induction amplitude in core leg;

selection of winding material.

Preliminary transformer calculation and determination of basic dimensions of the active part shall have to take into account given values of the rated parameters.

Calculation of HV and LV windings data:

selection of HV and LV winding type;

LV winding data calculation;

HV winding data calculation.

Determination of the SC parameters:

(a) SC losses — main and stray losses i0n windings; SC impedance (%).

Final magnetic system calculation. No-load parameters determination:

- package size and active sections of core leg and yoke;

- magnetic core system size;
- laminated steel weight;
- no-load losses;

Initial data for transformer calculation are the following:

- rated power;
- power frequency;
- phase number;
- rated line current of HV windings;
- rated line current of LV windings;
- winding connection diagram and vector group;
- SC impedance (%);
- rated current density in HV windings;
- rated current density in LV windings;
- normalized SC losses;
- normalized no-load losses.

Basic dimensions and electromagnetic parameters, determined by the general calculation of transformer design:

- windings height,
- core legs window height,
- interaxial distance between core legs,
- mean diameter of HV winding,
- radial size of HV winding,
- mean diameter of LV winding,
- radial size of LV winding,
- turn number of HV winding,
- turn number of LV winding,
- magnetic induction amplitude in core legs in no-load;
- weight of laminated steel of magnetic core.

As the modified parameters are set circumscribed circle diameter of the core leg dc and winding height h.

Let us consider the active part design as following [3].

Magnetic core has three-legs, flat-laminated, step lap core. Core legs and yoke sections are round stepped. Laminated steel brand and thickness – 3408, 0,30 mm. Density – $\gamma_{ст} = 7650$ kg/m^3.

The windings design. LV winding is internal, HV winding is external. The windings have the same height. The arrangement on the core leg is concentric. Windings wire materials are the following:

- copper (density – γwire = 8900 kg/m^3, specific electrical resistance at temperature 75°C – $\rho_{75} = 0{,}0210$ Ohm*mm^2/m);
- aluminium (density – $\gamma_{wire} = 2700$ kg/m^3, specific electrical
- resistance at temperature 75°C – $\rho_{75} = 0{,}0342$ Ohm*mm^2/m).
- Transformer calculation starts with determination of basic electrical
- values — power per phase and core leg, rated line and phase currents on HV and LV sides, rated phase voltages and currents [3]. Rated phase voltages and currents in HV and LV windings for a three-phase transformer are calculated on the basis of rated line currents and voltages; and winding connection diagram according to [3].

Calculation of transformer basic electrical values.

Rated phase voltage of HV winding of a three-phase transformer with windings in star connection:

$$U_{r1p} = U_{r1}/\sqrt{3};$$

with windings in delta connection:

$$U_{r1p} = U_{r1}.$$

Rated phase voltage of LV winding of a three-phase transformer with windings in star connection:

$$U_{2p} = U_{r2}/\sqrt{3};$$

with windings in delta connection:

$$U_{r2p} = U_{r2}.$$

Rated line current of HV winding of a three-phase transformer:

$$I_{r1} = S_r/(\sqrt{3}*U_{r1}). \quad (9.40)$$

Rated line current of LV winding of a three-phase transformer:

$$I_{r2} = S_r/(\sqrt{3}*U_{r2}).$$

Rated phase current of HV winding with windings in star-connection:

$$I_{r1p} = I_{r1};$$

with windings in delta connection:

$$I_{r1p} = I_{1r}/\sqrt{3}.$$

Rated phase current of LV winding with winding in star connection:

$$I_{r2p} = I_{r2};$$

with winding in delta connection:

$$I_{r2p} = I_{r2}/\sqrt{3}.$$

SC impedance of a transformer phase on the HV winding side:

$$Z_{sc} = (u_{sc}/100)U_{r1p}^2/S_{r.p}.$$

To calculate the active part basic dimensions and the parameters of the transformer, we apply the so called general calculation method [9, 10]. The general calculation method is based on the principles of the general theory of transformers and provides fairly simple and precise dependance between the specified ratings of the transformer, its geometric dimensions and the weight of active materials. It allows, without going into details, to calculate the basic dimensions of the magnetic core and windings, the weight of the active materials.

The features of the general calculation method are the following.

The calculation is performed without deepening into the small parts of the transformer. Simplifying assumptions are used. Some values are determined or calculated approximately and with significant changes in the initial calculation data they are changed insignificant. The real error in their approximate determination has a minimal effect on the calculation results. Only windings and a magnetic system are considered.

The windings are considered to be solid cylindrical. The transverse structure of the HV and LV windings is taken into account by the conductor filling factor of HV and LV windings, which is determined based on the experience of designing transformers, depending on the voltage class, the level of technology and thermal conditions.

It is assumed to be known in advance and stable for a transformer of a certain power and voltage class: the filling factor of the HV winding, the filling factor of LV winding, stray losses factor, the filling factor of circular cross-section of the core leg, the filling factor of round geometric cross section of the core leg.

9.9. The Calculation and Determination of the Basic Dimensions of the Active Part of the Transformer. Application of the General Method for the Transformer Calculation

The transformer is calculated in two stages: preliminary (estimated) calculation and final (detailed) calculation.

When choosing the values one should proceed with the accumulated experience of electromagnetic, electrical insulating and thermal calculations of power transformers of certain power and voltage classes and the achieved technological level in the company.

As a modifying parameter for transformer calculation is set circumscribed circle diameter of the core leg d_c with corresponding geometrical cross-section of the core leg

S_{geom} and active cross-section of the core leg S_c [10].

Rated phase voltages and rated phase currents of LV and HV windings are related directly with geometry of transformer active part.

Rated phase voltage of HV winding:

$$U_{r1p} = (\omega/\sqrt{2})w_1B_mS_c, \quad (9.41)$$

where w_1 — turn number of primary HV winding.

Rated phase voltages of LV winding:

$$w_1 = U_{r1p}\sqrt{2}/(\omega B_mS_c), \quad (9.42)$$

where w_2 — turn number of secondary LV winding.

This follows the turn number of HV winding:

$$w_1 = U_{r1p}\sqrt{2}/(\omega B_mS_c), \quad (9.43)$$

The turn number of LV winding:

$$w_2 = w_1*(U_{r2p}/Ur_{1p}). \quad (9.44)$$

Based on the results of leg section development the following parameters are clarified.

The clarified magnetic induction amplitude in magnetic core leg:

$$B_m = U_{r1p}\sqrt{2}/(\omega w_1S_c). \quad (9.45)$$

9.10. EQUATIONS OF CONNECTION OF THE BASIC DIMENSIONS OF THE TRANSFORMER WITH RATED VALUES

Basic dimensions and active part parameters are connected by means of electromagnetic power equation (equation of electromagnetic usage of active volume) (9.39) with the rated values and specific electromagnetic loads on active materials.

Equations of connection of the basic dimensions of the transformer with rated values by the SC impedance calculation.

SC impedance in percentage between HV and LV windings is calculated by the SC impedance calculation:

$$u_{sc} = (I_{r1p}Z_{sc}/U_{r1p})*100\% \ . \tag{9.46}$$

The rated values in this equation, specified by the technical task, are related with the active part geometry, accepted specific electromagnetic loads and other parameters in the following way.

Rated phase voltage of HV winding:

$$U_{r1p} = (\omega/\sqrt{2})w_1B_mS_c. \tag{9.47}$$

Magnetic moving force (MMF) of one winding:

$$w_1I_{r1p} = w_2I_{r2p} = \sqrt{2}S_{r.p}/(\omega B_mS_c). \tag{9.48}$$

If no-load current is negligible in transformer, then MMF of HV and LV windings arithmetically are equal to each other:

$$w_1I_{r1p} = w_2I_{r2p} \tag{9.49}$$

MMF of HV (LV) winding, with regard to (9.16):

$$w_1I_{r1p} = \sqrt{2}I_{r1p}U_{r1p}/(\omega B_mS_c) = \\ = \sqrt{2}S_{r.p}/(\omega B_mS_c). \tag{9.50}$$

MMF of windings are related with winding geometry in the following way.

Active cross-section area of LV winding turn:

$$S_{t2} = I_{r2p}/J_{r2}, \quad (9.51)$$

where J_{r2} — rated current density in LV winding. For example, for aluminium LV winding usualy is used [3] $J_{r2} = 1.50$ A/mm^2.

Active cross-section area of HV winding turn:

$$S_{t1} = I_{r1p}/J_{r1}, \quad (9.52)$$

where J_{r1} — rated current density in HV winding. For example, for aluminium HV winding usualy is used [9] $J_{r1} = 1.50$ A/mm^2.

Active cross-section area of LV winding with wire, with regard to (9.17):

$$S_{w2} = w_2S_{t2} = w_2I_{r2p}/J_{r2} = w_1I_{r2p}/J_{r2} = \sqrt{2}S_{r.p}/J_{r2}/(\omega B_mS_c). \quad (9.53)$$

Active cross-section area of HV winding with wire, with regard to (9.17):

$$S_{w1} = w_1S_{t1} = w_1I_{r1p}/J_{r1} = \sqrt{2}S_{r.p}/J_{r1}/(\omega B_mS_c) \quad (9.54)$$

Basic values by active part geometry calculation is namely a filling factor of winding cross-section. It connects electromagnetic and geometrical parameters of the transformer together.

The fundamental values are practically equal to 0.5, which considerably simplify the calculations.

Assuming that the HV and LV windings have the same height (Figure 9.5), we can obtain : the filling factor of HV winding cross-section:

$$k_{w1} = S_{w1}/(b_1h). \quad (9.55)$$

For example, $k_{w1} = 0.40 \ldots 0.60$.

That follows the radial size of HV winding, with regard to (9.20), (9.23), (9.25)), (9.56):

$$b_1 = S_{w1}/(k_{w1}h) = \sqrt{2}S_{r.p}/(\omega B_m S_c J_{r1} k_{w1} h). \quad (9.56)$$

The filling factor of LV winding cross-section:

$$k_{w2} = S_{w2}/(b_2 h). \quad (9.57)$$

For example, $k_{w2} = 0.40 \ldots 0.60$.

That follows the radial size of LV winding, with regard to (9.22), (9.24), (9.25), (9.58):

$$b_2 = S_{w2}/h = \sqrt{2}S_{r.p}/(\omega B_m S_c J_{r2} k_{w2} h). \quad (9.58)$$

9.11. Calculation and Determination of Basic Dimensions of the Active Part of the Transformer

The dimensions of laminated core stacks are calculated in such a way, that the stepped cross sectional figure fits in circumference of diameter dc. To core leg diameter value corresponds: n – core leg step number, b_{p1}, b_{p2}, … b_{pn} – set of width of laminated core stacks, S_{geom} – geometrical cross-section of core leg.

Cross-section shape of core leg in Figure 9.6.

The dimensions of laminated core stacks are the following (Figure 9.6):

thickness of the first stack:

$$a_{p1} = \sqrt{(\, d_c^2 - b_{p1}^2)}\,. \quad (9.59)$$

Thickness of the second stack:

$$a_{p2} = \sqrt{(\, (0.5d_c^2) - (0.5b_{p2})^2 - a_{p1}}. \quad (9.60)$$

Thickness of the nth stack:

$$a_{pn} = \sqrt{((0.5d_c)^2 - (0.5b_{pn}^2)} - a_{p1} - a_{p2} . \quad (9.61)$$

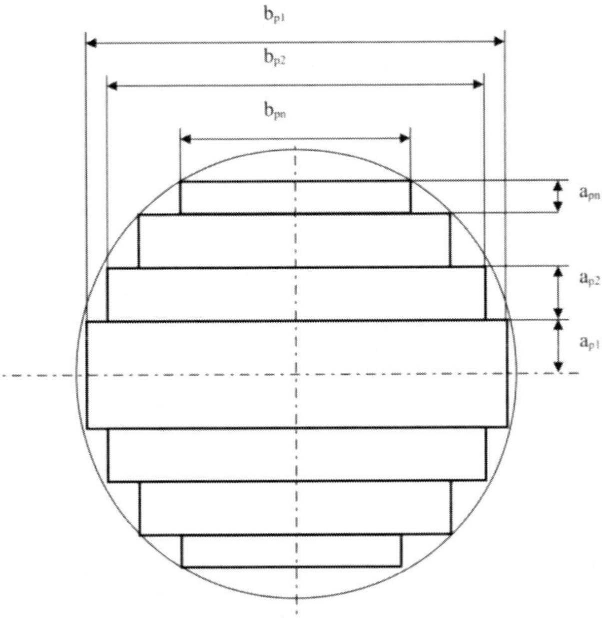

Figure 9.6. Cross-section of the core leg.

Geometrical cross-section of the core leg of set diameter is calculated in the followings way:

$$Sg_{eom} = b_{p1}a_{p1} + 2(b_{p2}a_{p2} + \dots + b_{pn}a_{pn}). \quad (9.62)$$

Magnetic core design is presented in Figure 9.7. Magnetic core basic dimensions and weight of magnetic core parts. Height of core window:

$$h_{wind} = h + \Delta_{y1} + \Delta_{y2}. \quad (9.63)$$

Interaxial distance:

$$L_{IA} = D_1 + b_1 + \Delta_{11}. \quad (9.64)$$

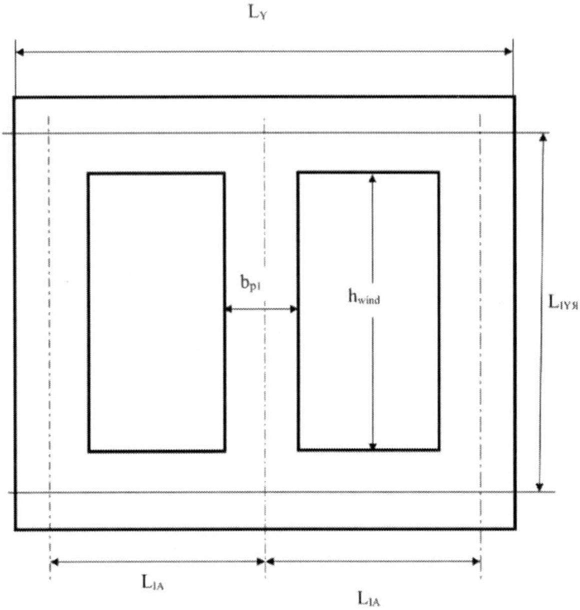

Figure 9.7. Magnetic core sizes (as per the 1st stack).

A three-leg flat magnetic core structurally is divided into core legs and yokes.

Yoke interaxial distance:

$$L_{IY} = h_{window} + b_{p1}. \tag{9.65}$$

Core leg length for the1st core stack:

$$L_{leg1} = L_{IY} - b_{p1}. \tag{9.66}$$

Core leg length for the 2nd core stack:

$$L_{leg2} = L_{IY} - b_{p2}. \tag{9.67}$$

Core leg length for nth core stack:

$$L_{legn} = L_{IY} - b_{pn}. \tag{9.68}$$

Table 9.6. Dependence of laminated steel characteristics on magnetic induction amplitude

Magnetic induction amplitude, Bm	Specific losses, Pspec		
	3408 – 0.30 mm	3407 – 0.30 mm	3406 – 0.30 mm
T	W/kg	W/kg	W/kg
1.60	0.961	1.014	1.069
1.62	0.994	1.052	1.112
1.64	1.032	1.094	1.156
1.66	1.076	1.139	1.201
1.68	1.125	1.187	1.251
1.70	1.178	1.238	1.307
1.72	1.235	1.293	1.369
1.74	1.294	1.351	1.433
1.76	1.355	1.411	1.498
1.78	1.419	1.479	1.564
1.80	1.485	1.549	1.632

Yoke length for the1st core stack:

$$L_{Y1} = 2L_{IA} + b_{п1}. \tag{9.69}$$

Yoke length for the 2nd core stack:

$$L_{Y2} = 2L_{IA} + b_{p2}. \tag{9.70}$$

Yoke length for nth core stack:

$$L_{Yn} = 2L_{IA} + b_{pn}. \tag{9.71}$$

Calculation of weight of magnetic core.
Core legs weight:

$$G_Y = 2\gamma_{cт}\, k_c(L_{Y1}b_{p1}a_{p1} + 2L_{Y2}b_{p2}a_{p2} + \ldots + 2L_{Yn}b_{pn}a_{pn}). \tag{9.72}$$

Weight of magnetic core yokes:

$$G_Y = 2\gamma_{cT}\, k_c(L_{Y1}b_{p1}a_{p1} + 2L_{Y2}b_{p2}a_{p2} + \ldots + 2L_{Yn}b_{pn}a_{pn}). \quad (9.73)$$

Weight of magnetic core laminated steel:

$$G_c = G_L + G_Y = \gamma_{cT}k_c m_p(L_{leg1}b_{p1}a_{p1} + 2L_{leg2}b_{p2}a_{p2} + \ldots + 2L_{legn}b_{pn}a_{pn}) + 2\gamma_{cT}\, k_c(L_{Y1}b_{p1}a_{p1} + 2L_{Y2}b_{p2}a_{p2} + \ldots + 2L_{Yn}b_{pn}a_{pn}). \quad (9.74)$$

9.12. Calculation of No-Load Losses

No-load losses are calculated in the following way. Specific losses in laminated steel $p_{spec} = f(B_m)$ are determined as per table dependence based on the magnetic induction amplitude. For laminated steel of 3408 - 0.30 mm, 3407 - 0.30 mm and 3406 - 0.30 mm type the dependence of specific losses (W/kg) on the magnetic induction amplitude in a table form are given in [8-13]. Specific losses in laminated steel are specified by empirical dependencies, based on the magnetic induction amplitude value $p_{spec} = f(B_m)$.

Dependence for laminated steel of 3408 - 0,30 mm, 3407 - 0,30 mm and 3406-0,30 mm type is presented in Table 9.4. Linear interpolation is applied for intermediate values calculation.

Dependence of laminated steel characteristics on magnetic induction amplitude is presented in the Table 9.6.

No-load losses are calculated by the following formula:

$$P_{no\text{-}load} = k_{tech}p_{spec}G_c, \quad (9.75)$$

where k_{tech} – technological factor, takes into account the increase of no-load losses due to technological factors, and is specified based on empirical data of transformers tests. Usually $k_{tech} = 1.28$ is used for step lap core

Recommended core leg diameters and plate width are given in Table 9.7.

Table 9.7. Normalized series of basic diameters and plate width

Diameter, mm	Stack number									Geom. section., cm^2
	№1	№2	№3	№4	№5	№6	№7	№8	№9	
95	90	70	50							59.0
104	100	90	70	50						74.1
116	110	90	70	50						92.95
128	120	100	90	70	50					115
145	140	120	100	70	50					149.0
164	160	140	120	100	70	50				193
178	170	160	140	110	90	70	50			230
198	190	170	150	130	110	90	70			285
218	210	190	170	140	110	90	70			344
254	250	240	210	180	160	140	120	90	70	475
275	270	240	210	190	170	150	130	110	90	553
295	290	270	240	210	190	170	140	110	90	641

9.13. SC Losses Calculation

SC losses are conventionally divided into main losses and stray losses. Main losses in oil transformer windings are calculated by known formula [10] of Ohmic losses.

Design temperature, to which SC losses are reduced, is 75°C [10].

For recalculation of reference specific electrical resistance of conductors to design temperature is used the following ratio [10]:

$$\rho_{(t2)}/\rho_{(t1)} = (T + t_1)/(T + t_2). \qquad (9.76)$$

where t_1 and t_2 are corresponding temperatures of conductor, $\rho_{(t1)}$ and $\rho_{(t2)}$ specific electrical resistance of conductor at t_1 и t_2 correspondingly, T is temperature, equal to 235°C for copper winding and 225°C for aluminium winding.

9.14. The Calculation of LV Winding Data

Wire length of LV winding:
The calculation of LV winding data.
Wire length of LV winding:

$$L_{wire2} = \pi D_2 w_2 \quad (9.77)$$

Direct current resistance of LV winding, with reference to (12):

$$R_{02} = \rho_{75} L_{wire2}/S_{t2} = \rho_{75} w_2 \pi D_2/S_{t2} = \rho_{75} w_2 \pi D_2 J_{r2}/I_{r2p}. \quad (9.78)$$

Wire weight of LV winding without insulation:

$$G_{wire2} = m_p \gamma_{wire} L_{wire2} S_{\text{в}t} = m_p \gamma_{wire} L_{wire2} S_{t2} I_{r2p}/J_{r2} = m_p \gamma_{wire} w_1 I_{r1p} \pi D_2 J_{r2} \quad (9.79)$$

Main losses in LV windings:

$$P_{main2} = m_p I_{r2p}{}^2 R_{02} = m_p I_{r2p}{}^2 \rho_{75} w_2 \pi D_2 J_{r2}/I_{r2p} = m_p J_2{}^2 \rho_{75} w_2 \pi D_2 S_{t2} = m_p S_{r.p} \rho_{75} \sqrt{2} \pi D_2 J_{r2}/(\omega B_m S_c). \quad (9.80)$$

9.15. The Calculation of HV Winding Data

Wire length of HV winding:

$$L_{wire1} = \pi D_1 w_1. \quad (9.81)$$

Direct current resistance of HV winding, with reference to (24):

$$R_{01} = \rho_{75} L_{wire1}/S_{t1} = \rho_{75} w_1 \pi D_1/St1 = \rho_{75} w_1 \pi D_1 J_{r1}/I_{r1p}. \quad (9.82)$$

Wire weight of HV winding without insulation:

$$G_{wire1} = m_p\gamma_{wire}L_{wire1}S_{t1} = m_p\gamma_{wire}L_{wire1}S_{t1}I_{r1}/J_{r1} = m_p\gamma_{wire}w_1I_{r1p}\pi D_1J_{r1} \quad (9.83)$$

Main losses in HV windings:

$$P_{main1} = m_pI_{r1p}{}^2R_{01} = = m_pI_{r1p}{}^2\rho_{75}w_1\pi D_1J_{r1}/I_{r1p} = = m_pS_{r.p}\rho_{75}\sqrt{2}\pi D_1J_{r1}/(\omega B_mS_c). \quad (9.84)$$

Main losses in transformer windings:

$$P_{main} = P_{main2} + P_{main1} = m_p(S_{r.p}\rho_{75}\sqrt{2}\pi(D_2J_{r2}+D_1J_{r1})/(\omega B_mS_c). \quad (9.85)$$

Total transformer SC losses:

$$P_{sc} = k_{add}P_{main} = k_{add}m_p(S_{r.p}\rho_{75}\sqrt{2}\pi(D_2J_{r2}+D_1J_{r1})/(\omega BmS_c). \quad (9.86)$$

Solution of nonlinear equations system, expressing the connection between the basic dimensions and transformer parameters, is specified by means of specific electromagnetic loads. System of nonlinear equations, expressing the connection between the basic dimensions and transformer parameters is the following:

$$
\left\{
\begin{array}{l}
w_1 I_{r1p} = \sqrt{2} S_{r.p}/(\omega B_m S_c); \\
Z_{sc} = \sqrt{X_{sc}^2 + R_{sc}^2}; \\
X_{sc} = \omega(k_\rho \mu_0 w_1^2 S_\sigma)/h; \\
R_{sc} = k_{add} \rho \pi \sqrt{2} S_{н.ф}(D_2/J_{r2} + D_1/J_{r1})/(\omega B_m S_c); \\
k_\rho = 1 - (b_1 + \Delta_{12} + b_2)/(\pi * h); \\
S_{\sigma sc} = \pi D_{12}(b2/3 \ + \Delta_{12} \ + b1/3 \ ); \\
D_2 = d_c + 2\Delta_c + b_2; \\
D_1 = D_2 + b_2 + 2\Delta_{12} + b_1; \\
b_1 = S_{r.p}(\omega B_m S_c J_{r1} k_{w1} h); \\
b_2 = S_{r.p}(\omega B_m S_c J_{r2} k_{w2} h); \\
P_{noload} = k_{tech} p_{spec} G_{cт}; \\
p_{spec} = f(B_m); \\
L_{IA} = D_1 + b_1 + \Delta_{11}; \\
h_{wind} = h + \Delta_{y1} + \Delta_{y2}; \\
L_{IY} = h_{wind} + b_{p1}; \\
L_{leg1} = L_{MY} - b_{p1}; \\
L_{leg2} = L_{MY} - b_{p2}; \\
L_{legn} = L_{IY} - b_{pn}; \\
L_{Y1} = 2L_{IA} + b_{p1}; \\
L_{Y2} = 2L_{IA} + b_{p2};
\end{array}
\right.
\quad (9.87)
$$

$$
\begin{array}{l}
L_{Yn} = 2L_{IA} + b_{pn}; \\
G_c = \gamma_{st} k_c[m_p(L_{cт1} b_{p1} a_{p1} + 2L_{cт2} b_{p2} a_{p2} + \ldots + 2L_{cтn} b_{pn} a_{pn}) + \\
+ 2(L_{Y1} b_{p1} a_{p1} + 2L_{Y2} b_{p2} a_{p2} + \ldots + 2L_{Yn} b_{pn} a_{pn})].
\end{array}
\quad (9.88)
$$

Solution of equation system (9.87) is entirely determined by means of specific electromagnetic loads.

9.16. An Example of the Transformer Calculation

Let us consider the transformer calculation example as per the proposed methodology. We take as a basis the example from [13-14]. Transformer 630 kVA, HV - 10000 V, LV - 400 V, windings connection diagram and vector group Y/Y_0 -0 [15-35].

Table 9.7. Main design data of transformer calculated ariants

Main data	Measuring unit	Variants		
		№ 1	**№ 2**	**№ 3**
Core leg diameter	mm	225	245	265
Core leg active cross-section area	m^2	0.03503	0.04150	0.04850
geometrical filling factor	-	0.917	0.917	0.917
winding height	mm	600	600	600
Interaxial distance	mm	435	435	435
Magnetic induction amplitude	T	1.649	1.430	1.191
Current density in LV winding	A/mm^2	2.406	2.406	2.406
Current density in HV winding	A/mm^2	2.477	2.477	2.477
LV winding filling factor	-	0.520	0.520	0.520
HV winding filling factor	-	0.461	0.461	0.461
Calculation results				
Turn number of LV winding	-	18	18	18
Turn number of HV winding	-	450	450	450
mean diameter of LV winding	mm	269.3	289.3	309.3
mean diameter of HV winding	mm	366.5	386.5	406.5
radial size of LV winding	mm	34.3	34.3	34.3
radial size of HV winding	mm	38.9	38.9	38.9
Wire weight of LV winding	kg	244.8	262.9	281.0
Wire weight of HV winding	kg	329.9	348.0	357.1
Weight of magnetic core	kg	1140	1367	1021
SC losses	W	9354	9913	10463
No-load losses	W	1491	1183	1021

We assume that LV winding – internal, HV winding – external. The windings have the same height. Windings arrangement on the core leg is concentric. Winding wire material is copper.

Insulation distances (Figure 9.5) are assumed to be equal for all variants: Δ_{y1} = 15 mm; Δ_{y2} = 15 mm; Δ_c = 5 mm; Δ_{12} = 10 mm. Main design data of transformer calculated variants are presented in Table 9.7 [35-47].

Magnetic core design data are given in Table 9.8.

Table 9.8. Magnetic core design data

Magnet core system					**Three-leg**
Laminated steel					3408, 0.30 mm
Diameter of magnetic core leg					198 mm
Magnetic induction amplitude					1.737 T
Core legs cross-section area					282.09 cm^2
Core legs active cross-section area					272.22 cm^2
Steel filling factor					0.96
Geometrical filling factor					0.920
Sizes of core stacks, mm	**Core stack**				**Plate number**
	number	**figure**	**width**	**thickness**	
	№1	2	190	28.0	90
	№2	2	170	23.0	74
	№3	2	150	13.7	44
	№4	2	130	9.95	32
	№5	2	110	7.35	24
	№6	2	90	6.2	20
	№7	2	70	4.4	14
	№8				
	№9				
Magnetic core window height					430 mm
Interaxial distance					380 mm
Laminated steel weight	net				662.1 kg
	gross				696.1 kg
Including:	core legs weight				662.1 kg
	yokes weight				696.1 kg

CONCLUSION

1. Splitting LV winding into lots of separate sections we can receive according to formula (9.4) any given relative content of HCH. And the lower the specified allowable HHF, for a larger number of section-stages one should subdivide CW, that is undesirable especially from the technological point of view. Therefore, to minimize the required CW sections number one should set whenever possible high allowable value of HCH factor, acceptable also for power system operation .
2. Using CSRT with smooth step-by-step regulation of operating current in modern power systems, allows effectively with minimum

HCH generation in network to compensate surplus reactive power at the consumers.

3. Using LC- filter of the most significant 5th harmonic, allows considerably to decrease regulation stage number and, moreover, to simplify CRT design. LC- filter can operate in this case at power frequency as a capacitor for compensation of inductive reactive power in the network.
4. The power equation of the transformer connects uniquely the active part geometry (windings height and diameters, core leg active section area) and specific electromagnetic loads (magnetic induction amplitude in core leg, rated current density in windings). Main electromagnetic parameters and dimension of the transformer active part may be obtained as a solution of power equation system.
5. The calculation methodology of the basic dimensions and parameters of the active part of a two-winding transformer by setting the specific electromagnetic loads (magnetic induction amplitude in the core leg, rated current density in the windings) is proposed. Defining the specific electromagnetic loads on the active part it is possible to evaluate transformer technical-commercial indices of the transformer. Moreover, increasing electromagnetic loads will lead to decrease of active materials application in transformer, and increase its no-load losses and SC losses.

ACKNOWLEDGMENTS

The authors are grateful to Aleksandrov G. N., academic, great electroenergetical scientist (his memory) and Department of Electrical and Electronic Apparatus of Saint-Petersburg State Technical University.

REFERENCES

[1] Reichert, K., Kauferle, L. (1974). Controllable reactor compensator for more expensive utilization of high voltage system. *Conferens Internationale des Grands Reseau Electrique (CIGRE).* 25 Session. Report, 32-07.

[2] Alexandrov, G.N., Albertinskiy, B.I., Shkuropat, I.A. (1994). Working principles of controlled shunt reactor of transformer type. *Elektrotechnics*, № 11 (in Russian).

[3] Alexandrov, G.N., Lunin, V.P. (2005). Controlled reactors. *Sankt-Peterburg, Centre of energetic personnel training*, (in Russian).

[4] Angot, A. (1967). *Mathematics for Electrical and Radio Engineers.* M.: Nauka. (in Russian).

[5] Didenko, V., Sirotin, D. (2012). Accurate measurement of resistance and inductance of Transformer windings. *XX IMEKO World Congress Metrology for Green Growth Selitember*, 9-14, Busan, Republic of Korea.

[6] Malinovskiy, V.N., Hoang, V.N., Muborakshoev, D.T. (2008). Control of turn deformation of power transformer windings// *Information Medium and Technology: Scientific Work. — Moscow.*, V. 3, 106-108 (in Russian).

[7] Khrennikov, A.Yu. (2016). *Diagnostics of Electrical Equipment Faults and Power Overhead Transmission Line Condition by Monitoring Systems (Smart Grid): Short-Circuit Testing of Power Transformers.* Nova Publishers, New York, USA, 165.

[8] Shkuropat, I. (2016). Transformers theory. Power equation and Arnold constant for two-winding transformers. *ELEKTRO.*, № 1, 17- 19 (in Russian).

[9] Pavlos, S.G. (2009). *Spotlight on modern transformer design.* London, NewYork: Springer-Verlag London limited.

[10] Tikhomirov, P.M. (1985). Transformers calculation. Moscow: *Energoatomizdat*, (in Russian).

[11] Leytes, L.V. (1981). Electromagnetic calculation of transformers and reactors. Moscow: *Energoatomizdat* (in Russian).

[12] National Standard GOST R 52719 – 2007. *Power transformers. General specifications* (in Russian).

[13] National Standard GOST 3484.1 – 88. *Power transformers. The methodology of electromagnetic tests* (in Russian).

[14] Dymkov, A.M. (1971). Calculation and design of the transformers. Moscow, *Vysshaya shkola* (in Russian).

[15] IEC 60076-2, Power transformers – Part 2: *Temperature rise for liquid-immersed transformers,* 2011, 100 p.

[16] IEC 60076-3, CEI 60076-3. *Power transformers – Part 3: Insulation levels, dielectric tests and external clearances in air*, 2013, 124 p.

[17] IEC 60076-5. *Power Transformers – Part 5: Ability to withstand short circuit*, 2006, 80 p.

[18] Verhoeven, B. (2016). Shunt reactor current tests with HV and UHV circuit breakers. More routine based switching of lower current can present significant concerns in distribution and transmission networks. *DNV GL's Quarterly Newsletter of Testing, Inspections & Certification*, https://www.dnvgl.com/article/shunt-reactor-current-tests-with-hv-and-uhv-circuit-breakers-76628.

[19] Bharadwaj Reddy, S. *Current Limiting Reactors and Short Circuit Currents*. https://instrumentationtools.com/current-limiting-reactors-and-short-circuit-currents/.

[20] Khrennikov, A.Yu, Shlegel, O.A., Lurie, S.I. Control and protection device of transformer windings against deformation at short circuits in service. Patent of Russian Federation, № 2063050. *Bulletin* № 18, (1996).

[21] Khrennikov, A.Yu. (1998). *Measuring device of the inductive transformer short-circuit impedance*, Patent of Russian Federation, № 96110718 A.

[22] Salama, M. A., Dudley, R.F., Gela, G. (1981). Calculation of steady-state surface stresses at winding ends of air-core power reactors. *IEEE Proceeding*.

[23] Fogelberg, T., Girgis, R.S. (1994). ABB power transformers - a result of merging different technologies with prospects for significant future

advancements, *International Simposium "Electrotechnics - 2010,"* Moscow, vol. 1, 56-59.

[24] Balakrishnan, V. (1989). A study of short-circuits in large power transformers, *Electrical India*, Vol. XXIX, № 4, 87-91.

[25] McNutt, W.J., Johnson, W.M., Nelson, R.A., Ayers, R.E. (1970). Power transformer short-circuit strength - requirements, design, and demonstration. *IEEE Trans. Power Appar. and Syst.*, 89, № 8.

[26] Zhang, F.J., Wu, Z.Y. (2009). Application actualities and development of short-circuit limiting technique for power system. *Guangdong Electric Power*, 22(2), 11-14.

[27] Lin, Y. (2010). Application of short-circuit current limitation technologies for 500kV power system in Guangdong Grid. *North China Electric Power, Mechanical & Electrical Engineering Techbology*, 39(9), 97-100.

[28] ABB_Australia. (2006). *ABB Capability Guide.* http://www04.abb.com/global/seitp/seitp202.nsf/c71c66c1f02e6575c125711f004660e6/57af7451ece31b4c482573ea001f8711/$FILE/AUABB%20Capability%20Guide_LR.pdf.

[29] Zhang, L., Guo, Q., Zeng, L.D., Dai, C. (2011). High-Capacity Explosive High-Speed Interrupted Mechanism for Independent Power System. *Journal of Naval Aeronautical and Astronautical University.*, 26 (5), 561-566.

[30] Han, G., Han, L., Wu, L. (2010). Application and development of methods on limiting power grid's short-circuit current. *Power System Protectionand Control*, 38(1), 131-144.

[31] He, Zh.Q., Gao,W.L. (2010). Analysis on restrictive measures against short-circuit current of 220kV Chongqing power grid. *Central China Electric Power*, 23(2), 31-34.

[32] Khrennikov, A.Yu., Shakaryan, Y.G., Dementyev, Y.A. (2013). Shortcurrent Testing Laboratories. Short-Circuit Performance of Power Transformers, Transformer Testing Experience. *International Journal of Automation and Control Engineering (IJACE)*, Volume 2, Issue 3, 120-127. http://www.seipub.org/ijace/AllIssues. aspx.

[33] Bertagnolli, G. (1998). Short-Circuit Duty of Power Transformers." *ABB Transformatori*, Legnano (Milano) – Italy, 1997.

[34] Janssen, A.L.J., te Paske, L.H., Smeets, R.P.P., Leufkens, P.P., Fogelberg, T. (2009).Years test Experience with short-circuit withstand capability of large power transformers. *CIGRÉ 2009, 6th Southern Africa Regional Conference*, Paper P501.

[35] Khrennikov, A.Yu. (2012). New "intellectual networks" (Smart Grid) for detecting electrical equipment faults, defects and weaknesses. *Smart Grid and Renewable Energy*, Vol. 3, No 3, 159-164. http://www.scirp.org/journal/sgre.

[36] Khrennikov, A.Yu. (2013). Fault Detection of Electrical Equipment. Diagnostic Methods. *International Journal of Automation and Control Engineering (IJACE)*, Vol. 2, No. 1, 19-27. http://www.seipub.org/ijace/Archive.aspx.

[37] Khrennikov, A.Yu. (2013). Smart Grid technologies for Detecting Electrical Equipment Faults, Defects and Weaknesses. *Workshop on Mathematical Modelling of Wave Phenomena with applications in the power industry*, Linnaeus University, Växjö, 23-24. http://www.lnu.se/mmwp.

[38] IEEE Std C57.12.90-1993, *Part II: IEEE Guide for Short-Circuit Testing of Distribution and Power Transformers.*

[39] Janssen, A.J.L., te Paske, L.H. (2000). Short-Circuit testing experience with large power transformers, *CIGRE SC12 Session, report 12–105.*

[40] Janssen, A.L.J., te Paske, L.H., Smeets, R.P.P. (2002). Short-circuit Tests on Large Transformers: Destructive or not? *TrafoTech*, Mumbai, India, 126-33.

[41] Janssen, A.L.J., te Paske, L.H., van der Linden, W.A., Smeets, R.P.P. (2001). Magnetic Saturation at Short-Circuit Tests on Power Transformers, *IEEE PES Summer Meeting*, paper 72.03.

[42] IEC Committee Draft 14/465/CD Annex B. *Theoretical Evaluation of the Ability to Withstand the Dynamic Effects of Short Circuit*, (2003).

[43] CIGRE WG12.05. An International Survey on Failures in Large Power Transformers in Service, *Electra*, (1983), 88, 21–37.

[44] Malewski, R., Anderson, W.I., Moore, W.J. (1980). Interlaboratory comparision of EHV shunt reactor loss measurements//*Transaction on Power Apparatus and System*, Vol. PAS-99, No. 4.

[45] DEP 33.65.40.31-Gen. Power transformers (Amendments/ supplements to IEC 60076-1 and IEC 60076-11). *Design and Engineering practice*, 2013, 31 p.

[46] European Standard. EN 50588-1. *Medium power transformers 50 Hz, with highest voltage for equipment not exceeding 36 kV - Part 1: General requirements*, 2015, 29 p.

[47] International Standard IEC 60076-10. *Power transformers – Part 10: Determination of sound levels*, 2001, 7 p.

Chapter 10

DIGITAL CURRENT TRANSFORMERS FOR MODERN DIGITAL SUBSTATIONS

A. Yu. Khrennikov, PhD*

ABSTRACT

The results of designing device for calculating current on digital current transformers, based on the Faraday Effect. The device calculates actual value of current, writes data to the non-volatile memory and sends status messages to the automated workstation (AWS). The element base of developed device was chosen, metrological calculations were made, the electrical circuit of device, the printed circuit board, and the algorithms of the device operation during the design process were developed.

Keywords: digital current transformer, current measurement, optical transformers, Faraday Effect, metrological calculation, circuit diagram

* Corresponding Author's E-mail: ak2390@inbox.ru.

10.1. Introduction

Substantial development of hardware and software management systems for use in electrical substations occurred since the start of development in the power industry projects of the automated process control system of the substation (ACS TP substation). High-voltage digital current and voltage transformers appeared. The primary and secondary electric network equipment with built-in communication ports are being developed. Microprocessor controllers are equipped with development tools, on the basis at which it is possible to create a reliable software and hardware complex of the substation.

The international standard IEC 61850 has been adopted. IEC 61850 regulates the presentation of data at substation as an object of automation, as well as protocols for digital data exchange between microprocessor intelligent electronic devices (IED) of substation, including control and management devices, relay protection and automation (RPA), emergency control automation (ECA), telemechanics, technological communication, electricity meters, etc.

All these factors create prerequisites for building a new generation of substation - a digital substation, in which the organization of all information flows in the solution of monitoring, analysis and control tasks is carried out in digital form.

10.2. The Advantages of Digital Substation in Comparison with a Traditional Substation (Given in Table 10.1)

The actuality of the construction of digital substations determines the design and development of digital devices intended for measuring, calculating and transmitting the parameters of current transformers.

Table 10.1. Comparison of the features of traditional and digital substations

Traditional	Digital
Losses in secondary circuits (for all devices are different)	Absence of loss of information
Multiple AC converters (in each device)	Single AC conversion (primary measurement)
Not synchronous measurements	Synchronous measurements
Great influence of electromagnetic effects	Absence of influence of electromagnetic effects
	Unlimited replication of information

It was not possible to eliminate the drawbacks inherent in current transformers despite the hundred years of experience in the development and operation of electromagnetic current transformers, and given that they are the main primary transducers for the purposes of relay protection and automation.

One of the significant drawbacks of the transformer is the saturation of the magnetic circuit during short circuits, which are accompanied by an aperiodic component. The use of open magnetic cores reduces saturation, but, other things being equal, reduces accuracy, both in current and in angle [1-3].

10.3. Optical Digital Current Transformers

Currently, optical current transformers on Faraday Effect are being actively advertised.

Despite the fact that the development of optical transformers has been under way for more than 40 years [2], the first industrial samples that appeared in recent years are too expensive, and can be used at high and extra high voltage (220 kV and more).

Practical interest is also represented by measuring current conversion based on galvanomagnetic effects used in magnetic field sensors that allow measuring not only alternating current but also direct current. These sensors can also measure alternating current with slowly decaying aperiodic components. The measured current creates a magnetic field, to which the sensor responds. Galvanomagnetic converters include Hall sensors, magnetoresistors, magnetodiodes and magnetotransistors.

Thus, the choice of the most promising types of primary current measurement converters for electrical engineering is actual.

A typical circuit of an optical current transformer (OCT) comprises a sensing element in the form of several turns of optical fiber placed in a rigid protective shell of a non-magnetic material: current head for OCT, and electron-optical unit (EOU) connected to a sensor element through an optical cross (Figure 10.1) [4-10].

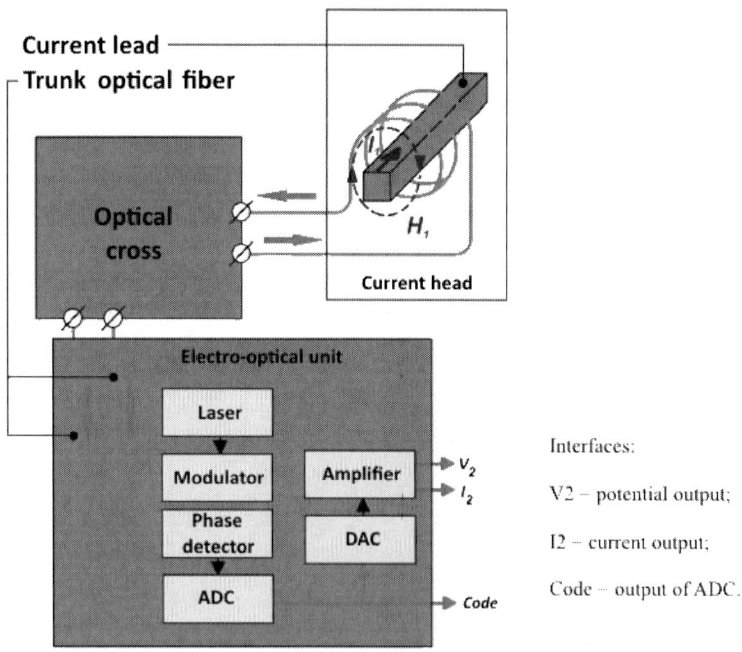

Figure 10.1. Simplified block scheme of an optical current transformer, where ADC - analog-to-digital converter.

The EOU generates on its optical output a monochromatic circularly polarized light signal with the help of an integrated laser and a modulator.

This signal is directed to the input of the sensing element over the optical fiber supporting the polarization. The polarization plane of the signal is subjected in the sensitive element to a magnetic field H_1 of the current I_1 corresponding to a rotation through the Faraday angle. The light signal is applied to the optical input of the EOU from the output of the sensing element, where an electrical measuring signal is formed from it on the phase detector.

Next, this signal is transmitted via an analog-to-digital converter (ADC) in the form of a digital code of a certain bit depth to the discrete interface output of the EOU and through a digital-to-analog converter (DAC) to the amplifier. Normalized analog output signals are formed in the form of voltage or current for issuance to the analog interface of an EOU at the amplifier. Thus, the measurement information can be obtained at the output of the EOU for further use.

The trunk optical fiber preserving the polarization and the cross unit (cable box), which connects it to the optical fiber of the sensitive element, are additionally used to connect the sensitive element. This is done in the case when, according to technological requirements, for example, at high-voltage substations, the sensitive element must be removed from the EOU [5-10].

10.4. Structural and Functional Scheme of the Device

The object of design in this work is a device used as an electronic-optical unit of modern current transformers based on the Faraday Effect. The generalized structural scheme of the device is shown in Figure 10.2.

Since the National Standard GOST R IEC 60044 [3] does not require mandatory presence of analog outputs on devices of this type, we confine ourselves to collecting and converting the measuring information.

The data from the digital measuring transformers, both optical and electronic, are converted into packets with the use of Merging Units, provided by the standard IEC 61850-9. The packets formed by the Merging Units are transmitted over the Ethernet network (process bus) to the level devices of the connection (controllers of automated process control system, relay protection & automatics, emergency control system, etc.) The sampling frequency of transmitted data is not worse than 80 points per period for relay protection & automatics and emergency control system and 256 points for the period for automated process control system, automated information-measuring system of commercial electricity metering, etc. Therefore, we need a high-speed microcontroller capable of processing in time and sending the processed data on the communication channels.

It is proposed to use the microcontroller (MC) of TMS 320 family of the Texas Instruments company - TMS320F2808 in this device. This MC family is designed for tasks where fast calculations are required, and for devices that do not require minimum power consumption. Intellectual peripherals, consisting of a wide range of digital and analog modules, do not use the central control processor (CCP) resource.

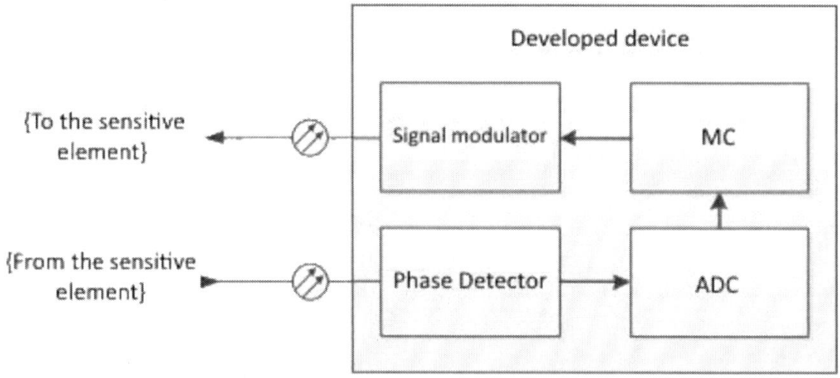

Figure 10.2. Generalized structural scheme of an optical CT.

The microcontrollers in this series contain most of the modules required to build the device. The interfaces for working with external devices are represented by the following set: UART, IrDA, SPI, I2C, USB, JTAG. It is possible to use the JTAG interface to debug a microcontroller. The converter will be able to connect to the programming unit via the JTAG interface.

There is ADC module - a 12-bit analog-digital converter of the conveyor type in the microcontroller TMS320F2808. Analog converter circuits include analog multiplexers with an external interface, sample and hold circuit, a converter core, voltage stabilizators and other auxiliary analog circuits. Digital circuits include a programmable device of cyclic control of the conversion (sequencer), registers of conversion results, interfaces to analog circuits and other modules on the chip.

The ADC sequencer can be configured in automatically mode a series of conversions. This means that every time the ADC receives the "start conversion" signal, it can perform plural conversions automatically. It is possible to choose any of the 16 input channels available through the analog multiplexer for each of them. After the conversion is completed, the digital value of the signal from the selected channel will be stored in the corresponding result register.

As a signal modulator, a laser light emitting diode of firm Laser Components will be used. It is possible to choose the model FBLD-785-1.60-FC-6Pin (Fiber-coupled High Power Laser Diode) among the extensive list of manufactured products. This product is unique in that it contains both an optical radiator and an optical signal receiver, and also comes with an optical fiber preinstalled in the housing of the product. This procedure is performed on high-tech equipment that allows to get rid of the errors associated with the loss of the light beam when the wrong pairing of optical fiber and light emitting diode. In addition, this product also contains an optical signal receiver, which is a special high-precision photodiode. The emitter and the receiver are in the same housing, which positively affects the overall dimensions of the device during design.

It is necessary to use a phase detector that will compare the senting signal with the received signal in order to obtain results from the application of the Faraday Effect. Analog Devices AD8302 solution (RF/IF Gain and Phase Detector) was chosen as such a component. This product can be used as a phase detector paired with photodiodes. The source signal and the signal from the photodiode are connected to the input. We get a voltage corresponding to the deviation of the phase of the received signal from the sent signal at the output of the product. This voltage will be applied to the ADC.

OLED display Bolymin BL1002A was selected as a means of interaction with the device. This character display is quite suitable for displaying the device status, the measured current value or the service information for interaction with the device operator. This display is powered by a 5V circuit. The manufacturer guarantees a service life of 10,000 hours. The operating temperature range is from -40 to 85C.

The main purpose of the memory chip in the developed device is a buffer for storing the transmitted information. As the buffer is filled, the buffer will be cleared, and the data will be transferred to the AWS. In addition, the accumulated data can be used for compilation reporting and calculations of statistical values. In addition, the accumulated data can be used to compile reports and calculate statistical values. Also, in case of abnormal situations, the device will keep data in the memory device, indicating the time and date the fault was detected.

We will assume that the volume of one data packet obtained by processing input electrical quantities is approximately 200 bytes. In this case, the use of a buffer memory with a capacity of 16 Kbit will completely satisfy our needs.

There are various technologies for implementing non-volatile memory, such as EEPROM, Flash, FRAM and so on. It is necessary to take into account the reliability requirements of the device when choosing the appropriate technology for a specific task. In the context of the tasks implemented programmatically, they imply:

- high memory speed;
- a sufficiently large number of rewriting cycles (it will allow to increase the service life of the device without the need for servicing, and also to minimize data losses in case of power outages).

It was decided to use memory of the FRAM type, since this type is most advantageous for the write speed, the time of resource (the number of read-write cycles) and the current consumption after comparing the characteristics of memory chips of different types. Since the memory types built into the microcontroller have worse characteristics than the FRAM memory, the use of an external chips buffer memory FM25L16.

According to the standards, this device should be able to be powered both from the 220 V network and from an autonomous power source.

The power supply of the device is proposed to be organized as follows: from the external power supply to the device is supplied 12V. Then, it is lowered first to 5V by two linear regulators, for powering the analog part of the device, then to 3.3V for powering the digital part. As regulators, linear regulators of the company Analog devices, model ADP3331, were chosen. Own consumption of the regulator does not exceed 34 μA.

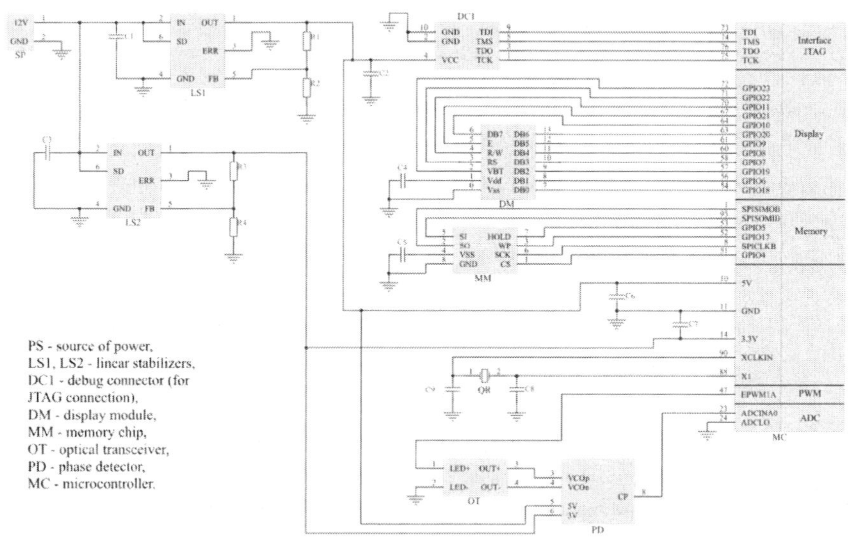

Figure 10.3. Functional scheme of the device.

As autonomous power supply is suggested to use the battery H&T YT12A-BS with the following characteristics: voltage 12V, capacity 10 A/h, temperature range from -30 to +50°C.

The functional scheme of the designed converter is shown in Figure 10.3. The design was carried out in Altium Designer 14 [3-10].

After passing through the light beam of the current sensor based on the Faraday Effect, we have two light signals: the original signal and changed signal, after passing through the magnetic field, created by the current in the current conductor. The deviation of the light beam is fixed by the phase detector, converted to voltage and passes through the analog-to-digital conversion. The obtained value is used to calculate the current. Current measurement error consists of the error of the current sensor, the error of conversion of the phase detector, the error of the ADC. The error of the current sensor in this calculation is not considered, since the current head is outside the device.

10.5. Error of the Device

Thus, the inaccuracy of the device consists of the inaccuracy of the optical transceiver, the inaccuracy in converting the phase detector and the inaccuracy of the ADC. The manufacturer stated that the inaccuracy of the transceiver δ_{tr} does not exceed 0.05% over the entire temperature range.

The nonlinearity of the phase detector conversion is additive and does not exceed 0.7° with a phase difference equal 90°. The voltage at the output of the phase detector varies linearly, and with a change in the phase shift from 0° to 90° varies between 1.8 V and 0.9 V.

The inaccuracy of the ADC is due to:

- integral nonlinearity;
- quantization error and noise;
- an offset of zero;
- full scale error;
- instability of the gain of the programmable amplifier.

The error due to the integral nonlinearity has an additive character and is equal to:

$$\Delta_{int} = 10^{-6} \cdot FSR_{max} = 10^{-6} \cdot \frac{1{,}024 \cdot REFIN1}{Gain} = 10^{-6} \cdot \frac{1{,}024 \cdot 2{,}50V}{1} = 2{,}56\mu V.$$

The error due to the zero offset is calculated as follows:

$$\Delta_{off,ADD} = \Delta_{off} + \Delta_{off,dr} \cdot \Delta t = 1\mu V + (10nV/°C \cdot 40°C) = 1{,}4\ \mu V.$$

The error of the full scale is multiplicative nature and is maximum equal:

$$\Delta_{FS} = \pm 10\ \mu V.$$

According to the ADC documentation, Quadratic Deviation of noise is equal

$$\sigma = 0{,}065\mu V$$

Therefore, the error due to the error of the quantization and noise is equal:

$$\Delta_n = 3 \cdot \sigma = 3 \cdot 0{,}065\ \mu V = 0{,}195\ \mu V.$$

The error due to the instability of gain factor of the programmable amplifier is equal:

$$\Delta K_{pamp,FS} = K_{pamp} \cdot \Delta t \cdot U_{FS} = 3 \cdot 10^{-6} \frac{1}{°C} \cdot 40°C \cdot 2{,}5\ V = 0{,}3\ mV.$$

The total additive component of the ADC error:

$$\Delta_{ADD,\Sigma} = \Delta_{int} + \Delta_{off,ADD} + \Delta_n = 2{,}56\mu V + 1{,}4\mu V + 0{,}195\mu V = 4{,}155\ \mu V.$$

The total multiplicative component of the ADC error:

$$\Delta_{FS,\Sigma} = \Delta_{FS} + \Delta K_{pamp,FS} = 10\mu V + 0{,}3mV = 0{,}31\ mV.$$

Below is the calculation of the error of the device, brought to the input of the phase detector.

$$d_{calc} = \delta_{tr} + \delta_{pd} + \frac{\Delta_{ADD,\Sigma}}{U_{FS}} \cdot 100\% = 0{,}05\% + 0{,}7\% + \frac{4{,}155\mu V}{2{,}5V} \cdot 100\% = 0{,}751\%$$

$$(c-d)_{calc} = \frac{\Delta_{FS,\Sigma}}{U_{FS}} \cdot 100\% = \frac{0{,}31mV}{2{,}5V} \cdot 100\% = 0{,}0124\%$$

$$c_{calc} = d_{calc} + (c-d)_{calc} = 0{,}751\% + 0{,}0124\% = 0{,}7634\%$$

Inaccuracy of the device (at the input) is determined by the formula:

$$\delta_{device} = c + d \cdot \left(\frac{\Delta\Phi_{FS}}{\Delta\Phi_{meas}} - 1\right) \approx 1{,}5\%,$$

where $\Delta\Phi_{FS}$ is the final value of the phase difference;

$\Delta\Phi_{meas}$ is the measured value of the phase difference.

The main contribution to the additive component of the error of the device is made by the phase detector, which can be seen from the results of the calculations. It is necessary to take into account the material from which the optical fiber is made and the number of turns around the conductor when calculating the error of current measurement. For example, the additive component of the measurement error of the current will decreased by a factor of 100 compared to the inaccuracy of this device, with N = 100 turns, and when recalculating it will be equal

$$\delta_1 = 0{,}02\% + 0{,}008\% \cdot \left(\frac{\Delta I_{FS}}{\Delta I_{meas}} - 1\right) \approx 0{,}030\%,$$

where ΔI_{FS} is the final value of the current;

ΔI_{meas} is the measured value of the current.

This error is small and commensurate with modern measuring current transformers. Table 2 shows examples of fiber optic current sensors from different manufacturers with accuracy classes.

Table 10.2. Accuracy classes of some current transformers

Firm	**NxtPhase**	**PowerSense**	**OptiSense**	**FieldMetrics**	**ABB**	**Airak**	
Accuracy class,%	0.25	2	0.2	0.2	0.2	1	0.25

10.6. EXPERIMENTAL POLYGON "DIGITAL SUBSTATION"

The development of digital substations is actively involved today in Russia. There is an experimental polygon, the first in Russia "Digital substation," on the territory of JSC "Scientific and Technical Center of FGC UES" [11] (Figure 10.4).

The experimental polygon consists of two parts - substation and laboratory. The substation part includes the substation control center, measuring optical current and voltage transformers, field converters of analog and discrete signals of external design. The laboratory part is a set of software and hardware test complexes (RTDS, OMICRON CMC 256 plus, RETOM-61850).

Optical current and voltage transformers of the NxVCT type and field current and voltage converters of the Merging Unit type are installed on the 110 kV and 10 kV side in an open switchgear of a digital substation. Communication environment for secondary equipment was established on the basis of digital switches RSG 2100 and RSG 2288. DPM-121 type universal multifunctional intelligent electronic devices (IEDs) are used to measure the effective values of current and voltage, active and reactive power and energy, and power quality parameters. RET-670 power transformer differential protection, REC-670 connection controller, busbar protection SEL-421-SV prototype and SEL-421-SV backup connection controller are used in the relay circuit. The SCADA-system KOTMI-2010 is used to control the substation, which communicates with IED without the use of OPC technology for the protocol driver IEC 61850-8-1. To control the substation is the SCADA-system KOTMI-2010 used, which through the IEC 61850-8-1 protocol driver, communicates with the IED without the use of OPC technology.

The experimental polygon allows to conduct research on building the architecture of the environment communication for various types of substations. Technical requirements will be formed for all types of equipment for use in digital substations in the process of research. Also, an experienced polygon will allow personnel of branch organizations to familiarize themselves with the new technology and receive training. Based on the results of the tests, it will be possible to determine the prospects for the development of digital substations, as well as their implementation in the Russian electric power industry.

Figure 10.4. "Digital substation" of JSC "STC of FGC UES."

Since the substation part of the digital substation already includes optical current and voltage transformers, it is currently possible to implement the developed device as part of the commercial electricity metering system [11].

The use of digital substations will significantly reduce the costs of transmission, conversion and distribution of electricity by improving the quality and reliability of the operation and operation of substations, as well as reducing the cost of maintenance of substations and ensuring the economic safety of substations.

ACKNOWLEDGMENTS

Dr. Sergej G. Popov, the head of digital substation of JSC "Scientific and Technical Center of FGC UES"; Mr. Alexander Skorblyuk, Master of Science, Russia, MPEI, greatly acknowledged for supporting this study. Cooperation of Universities and Innovation Development, Doctoral School project "Complex diagnostic modeling of technical parameters of power transformer-reactor electrical equipment condition," 2009, has made publishing of this article possible.

CONCLUSION

1. An electron-optical block of a current transformer based on the Faraday Effect is presented.
2. Metrological calculation of measurements of instantaneous value of current made. The inaccuracy of measuring the current at the input is not more than 0.030%.
3. It is possible to implement the device as part of the system of commercial electricity metering for the "Digital substation" of JSC "STC FGC UES."
4. There is first an experimental polygon "Digital substation" in Russia.

References

[1] *EC 61869-9:2016 | IEC Webstore | LVDC*. webstore.iec.ch. Retrieved 2018-07-12.

[2] Facchin, C. (2017). We are bridging the gap. Enabling the ABB digital substation. *ABB brochure.* https://search-ext.abb.com/library/Download.aspx?DocumentID=4CAE000291&LanguageCode=en&DocumentPartId=LowRes&Action=Launch.

[3] McFadyen, S. *How a Digital Substation Works*. https://myelectrical.com/notes/entryid/245/how-a-digital-substation-works.

[4] EcoStruxure™ Substation Operation. Digital Substation. *Schneider Electric*. https://www.schneider-electric.ru/ru/work/products/product-launch/digital-substation/.

[5] Gavrichev, V.D., Dmitriev, A.L. (2013). Fiber-optic magnetic field sensors. *Study Guide,* St. Petersburg (in Russian).

[6] National Standard GOST R IEC 60044-8-2010 *Measuring transformers. Part 8. Electronic current transformers,* Date of introduction 2012-07-01 (in Russian).

[7] *Digital substation*, NR Electric Co., Ltd, access date: March 2015

[8] *Optical measuring transducers*. www.tektologic.ru. (in Russian).

[9] Vlasov, M., Serdtsev, A. (2006). High-Voltage Optical Converters for Systems for Measuring and Analyzing the Quality of Electric Energy. *Energorynok,* No. 10 (in Russian).

[10] Morzhin, Yu. I., Popov, S. G., Rumyantsev, A. A., Il'in, M. D. (2014). The experimental test site of JSC NTC FGC UES is the first in Russia "Digital substation" 110 kV using the standard IEC 61850 (-8-1 and -9-2LE). *Energy of the Unified Network*, No. 3 (in Russian).

[11] *Digital Substation*. http://digitalsubstation.com/en/about/.

About the Authors

Alexander Yu. Khrennikov was born in Bratsk, Russia, in 1964. He received Philosophy Doctor Degree in Electrical Engineering from Samara City University of Technology in 2009 in the field of diagnostic modeling of technical parameters of power transformer-reactor electrical equipment.

Now, he works as Professor at the Moscow Energy Institute, Russia. The author has more than 240 scientific and technical publications. His main research interests concentrate in the field of Transformer Short-circuit testing, Transformer winding fault diagnostic, Frequency Response Analysis, Smart Grid and Information-measuring systems. He is distinguished CIGRE member.

Alexei A. Kuvshinov was born in Togliatti, Russia, in 1953. He is currently a Professor at Togliatti State University. He received his Philosophy Doctor Degree in Electrical Engineering from Ulyanovsk State University of Technology in 2004. The author has more than 200 scientific and technical publications. His research interests include converting equipment, logical-algebraic methods for analyzing and synthesizing conversion devices, electric power industry, including electrodynamic testing of power transformers, reactive power compensation, the influence of geomagnetic disturbances on the operation modes of electric power systems.

Igor A.Shkuropat was born in Togliatti, Russia, in 1964. He is currently Chief Specialist, Electroshield-TM, Samara. He received a Philosophy Doctoral Degree in Electrical Engineering from Saint-Petersburg State University in 2001. His main research interests concentrate in the field of controlled reactor-transformers, calculation of parameters of transformer active part, electromagnetic calculations of power transformers and reactors.

INDEX

A

B

E

F

G

H

I

L

M

N

O

U

V

W

X

Y

Z

Analysis and Control of the Chaotic Behavior in a Multi-Cell DC/DC Buck Converter

Author: Karama Koubaâ

Series: Electrical Engineering Developments

Book Description: In this book, we deal with the case of an N-cell DC/DC buck converter, and we present a systematic method to control the voltage across the flying capacitors and the current through the load that can be generalized from a two-cell converter to an N-cell converter.

Hardcover ISBN: 978-1-53613-998-3
Retail Price: $160

Photovoltaic Systems: Design, Performance and Applications

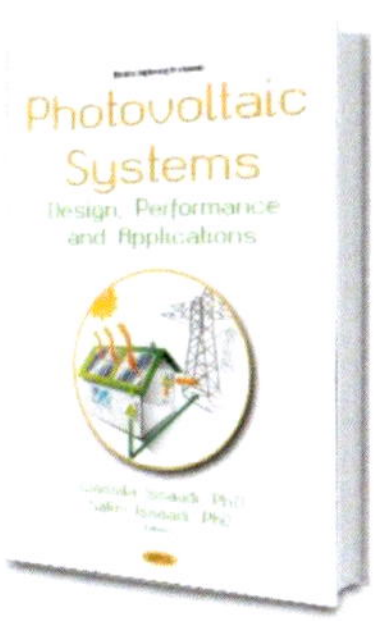

Editors: Wassila Issaadi, Ph.D. and Salim Issaadi

Series: Electrical Engineering Developments

Book Description: Presenting a complete guide for the planning, design and implementation of solar PV systems for photovoltaic (PV) applications, this book features analyses based on the authors' own laboratory testing as well as their experiences in the field.

Hardcover ISBN: 978-1-53613-646-3
Retail Price: $230